OUR WORLD IS BURNING

My Views on Mindful Engagement

Ian Prattis

Manor House

Ian Prattis / *Our World is Burning*

Library and Archives Canada Cataloguing in Publication

Prattis, J. Ian, 1942-, author
Our world is burning : my views on mindful engagement /
Ian Prattis.

 ISBN 978-1-988058-27-6 (hardcover).
--ISBN 978-1-988058-24-5 (softcover)

1. Environmental protection--Religious aspects.
2. Environmental protection--Religious aspects--Buddhism.
3. Human ecology--Religious aspects.
4. Human ecology--Religious aspects--Buddhism.
5. Environmental protection--Citizen participation.
I. Title.

BL65.E36P735 2017   201'.77   C2017-906350-2

Printed and bound in Canada / First Edition.

Cover Design-layout / Interior- layout: Michael Davie
Cover illustration: Johan Swanepoel / Shutterstock
Interior Text Edit: Meghan Negrijn and Michael B. Davie
192 pages. All rights reserved.
Published Oct. 15, 2017
Manor House Publishing Inc.
452 Cottingham Crescent, Ancaster, ON, L9G 3V6
www.manor-house.biz
(905) 648-2193

"This project has been made possible [in part] by the Government of Canada. «
Ce projet a été rendu possible [en partie] grâce au gouvernement du Canada."

Funded by the Government of Canada
Financé par le gouvernement du Canada   Canadä

Choose Love, Decency – not the Ego

PUBLICATIONS BY DR. IAN PRATTIS

* New Directions in Economic Anthropology. Special Edition of the Canadian Review of Sociology and Anthropology, August 1973

- Leadership and Ethics

- Reflections: The Anthropological Muse

- Anthropology at the Edge: Essays on Culture, Symbol and Consciousness

- The Essential Spiral: Ecology and Consciousness After 9/11

- Failsafe: Saving The Earth From Ourselves

- Earth My Body, Water My Blood

- Song of Silence

- Redemption

- Trailing Sky Six Feathers: One Man's Journey with His Muse

- New Planet New World

- Nine eBooks on Amazon Kindle

    2 CD's and 2 DVD's

- 4 films

- 8 Professional Honors, 3 book awards

- 10 Scientific and Technical reports

- 100 professional articles/chapters/book reviews published

- 26 Electronic publications of television courses broadcast at Carleton University and on TVO

- 50 articles in Pine Gate – Online Buddhist Journal; 200 articles in newspapers, community magazines

PRAISE FOR *OUR WORLD IS BURNING*:

- **Allan Green,** Spiritual Facilitator:

  I have eternal admiration for the wisdom, abilities and vision of Ian Prattis, amazed by his continuous supply of such great works. *Our World is Burning* is so poignant and necessary for the state of our world. This new book celebrates one of the great visionaries of our times.

- **Barbara A. White,** MA:

  Ian writes with a monastic lyricism that is disarmingly invigorating, ideologically and pragmatically (that is the point) empowering. Ian is a "meme" catcher, a "meme" weaver of cross–cultural and pan-generational challenges and continuities. Like Franklin with the fabled key on his high flying kite, Ian is standing in the rain capturing lightning bolts. Gird yourself! *Our World is Burning* is a lightning bolt and it will singe incredulity and cynicism.

- **Laurence Overmire,** author of ***The One Idea That Saves The World***:

  Our World is Burning is an inspiring and informative read. As the title suggests, we are living in challenging and perilous times. Ian Prattis offers us valuable insight, wisdom and perspective in finding our way to a healthier world, one based on compassion and commitment, mindful of how everything we do impacts the whole.

5

- **Marianna van de Lagemaat,** Herbalist Farmer:

The words in these essays touch my heart deeply. For instance: "I don't want to grow up and live in a world that is burning." From the heartfelt cry of a terrified nine-year-old boy, Ian demonstrates in his thoughtful, gentle way how we can, through an awakened awareness, change ourselves as individuals, think mindfully, sustainably and globally as interconnected communities and thus heal our Earth and restore our humanity. In Ian's words, "The task before us in the 21st century is to step out as spiritual warriors."

- **Koozma J. Tarasoff,** Anthropologist, Peace Activist, Author and Photographer:

Ian Prattis' new book is an urgent call for action in our troubled world. Environmental pollution, wars, violence, greed, ego worship and crass materialism are issues that urgently need to be resolved for the health of our Planet Earth and its inhabitants. Indeed, we need to release our bad thoughts to the soil and become informed. This is a book for the new generation who also need to nurture a respectful relationship to Mother Earth. Bravo, Ian, for helping to bring the tipping points of our consciousness closer to a critical mass for radical change. You are a Master story teller and healer with a vision for a new world.

- **Dawn James,** Conscious Living Advocate:

  Ian Prattis provides a conscious-raising framework for responsible living. *Our World is Burning* also challenges one to become a leader for change instead of a passive bystander. For this personal transformation to occur we must examine our values, behaviours and consumption patterns. One of the aspects I love about *Our World is Burning* is Ian's ability to describe our global crisis through different lenses, including political, environmental, cultural, economics and consciousness. This book shows us thought-provoking evidence that we are, in fact, our environment and therefore we are responsible to change it.

- **Susan Taylor Meehan,** author of *Maggie's Choice*:

  Our World is Burning is both a cry from the heart and a call to action. In clear, compelling prose, Dr. Ian Prattis, Zen teacher, ecologist and peace activist, outlines the urgent challenge of climate change and the prevailing attitudes that have enabled it to threaten life on planet Earth. He writes movingly of his own journey towards enlightenment to illustrate his basic thesis: that we cannot heal our planet until we heal ourselves. Drawing from the wisdom of Indian gurus, First Nations shamans, Buddhist teachers and activists, and the probing questions of his own grandchildren, Dr. Prattis shows that we need to undergo a radical transformation of the mind if we are to respond to our burning world. His call to action is for every individual to take up the challenge in mindfulness, love and compassion, and to build the kind of world that nourishes and sustains us all.

- **Michael B. Macdonald,** Film-maker, Associate Professor at MacEwan University:

I was introduced to Ian Prattis when he was founding Friends for Peace. I was deeply engaged in a personal struggle to articulate a way of living that was committed to anti-oppression, anti-war and anti-inequality. Ian helped me understand that fighting against injustices needs to be built around the cultivation of new ways of being, being together and being on the planet. I began to learn about the importance of consciousness in the struggle to want less, want peace, love. Ian's vision is complex and important. His ideas are rich and rewarding if you take the time to sit thoughtfully with them. Ian is committed to changing the world, and unlike those who may wait for a new world to come, Ian has developed practices to bring a new world into being. My hope is that this book will be read widely, and ideally, in community.

- **Gayle Crosmaz-Brown,** Master Drum Artisan & Spiritual Activator:

Ian's writing style keeps one motivated to keep turning the pages wanting to know more. His passion for sharing his insights and growth has no bounds, and triggers others to take action. May all who invest their time absorbing these pages find it in their own hearts to live the example being created within his prose. Ian is the stone being tossed into the waters of life. Let his ripple be felt on all shores.

- **Jacqueline Schoemaker Holmes,** PhD:

  Dr. Ian Prattis is a visionary and leader in the world of engaged Buddhism. Ian is the epitome of practicing what one preaches and encourages – in fact, inspires – others to do the same. This book reads like an invitation – the same kind of invitation he extends to those in search of a community and who find the gentle embrace of the Pine Gate Mindfulness Community. Ian provides what we need in troubled times – clear guidance, practical steps to take, and a warm and open hand in a world that so many fear is becoming too cold and distant. Ian's writing offers a welcome to his hearth where he gifts us with the impossibly perfect teachings of interconnection and heart opening. In this book, Ian makes an excellent contribution to existing commentary on world change and inspires action through the wisdom of his engaging story-telling.

- **Jana Begovic,** author of *Poisonous Whispers*:

  Ian Prattis' writings reflect the essence of his character. A steward of Gaia, in his opening chapter "Our World Is Burning" Prattis engages in a dialogue with a nine-year old boy who is terrified of growing up in a world that will burn up. With touching gentleness, Prattis assuages the boy's fears and paints an image of the role the boy can assume in contributing to the forces of good in the world. The essay highlights what most of us are aware of - the devastation inflicted by humans on planet Earth. Prattis shines the light on the path of mindful living by outlining a series of steps we could all adopt in our effort to reduce the negative imprint on Earth. The thread of mindfulness in the tapestry of Prattis' essays encompasses the acts of transferring knowledge to others. In "Punk Palace" Prattis describes how giving his son the gift of mindfulness, he saves him from the clutches of drug addiction. Prattis treads a terra nova and offers a tale of a dad's love, patience and devotion. His other essays exude profoundly inspirational messages and sound the alarm bell, but also light the torch of hope, and possible redemption of a landscape of potentially apocalyptic darkness. This book is a gift.

- **Jim Ebaugh,** Founder of Water in the Wave Communities and Engaged:

Dr. Prattis has been a voice, a teacher, a passionate advocate for the earth and all her species for decades and long before the word eco-dharma entered our vocabulary. His books transcend time and space. Ian Prattis was at the forefront of awakening to the threat that climate change posed to our home and our mother – the Earth. What once was ridiculed as extremist is now evident in everyday news, as we hurtle towards a tipping point, which has already been born, Ian's books are a creative, imaginative read as we struggle to find a new paradigm for our culture – away from rampant, unfettered consumerism and global corporate oligarchies demanding ever increasing short term profits at the expense of earth and all her species. Prattis leads the way in this collection of essays.

- **Ute H. Webb:**

  Ian has dedicated his life to engaged Buddhism, leading by example and guiding us to become guardians of the earth, always challenging us to reach beyond our comfort zone. Drawing on his deep respect for the teachings of Zen Master Thich Nhat Hanh, his long training as a Shaman, and his years spent as a Yogi in India, his knowledge, wisdom and reflections are timely and ever so pressing. He offers, and is, mindful engagement at its finest. In his brilliant conversation with young James in Chapter One he lays out the path we must choose. Failure to do so is summarized in his excellent Rant from the Future – Chapter Two. Rock on Ian!

- **Peggy Lehmann,** author and medium:

  Ian's essays on mindful engagement are an overview of a lifetime's work that started with a younger version of himself and a goal of saving the world. Through his books *Redemption*, *Trailing Sky Six Feathers*, and *New Planet, New World* readers saw glimpses of the man and his message both evolving and growing to new levels of spiritual understanding. At a time on earth when hope is badly needed, Ian's writings have universal appeal, assuring us that a better world is possible and that each of us must contribute to its creation. Enlightening!

- **Tonya Pomerantz,** creator, Puddle Jump Coaching:

Ian Prattis' writings on mindful engagement are a crucial read. Prattis offers the reader a look at the future and climate change through the eyes of a nine year old boy. This allows readers to explore the topic with a different filter. I wasn't a middle aged woman as I read this. I was transported to childhood. Prattis focuses on the possibilities that exist if we, as one world, can choose a positive way forward. Open, honest, down-to-earth and authentic, Prattis shares his stories of family and community. His style of writing is inviting; welcoming the reader on a global journey filled with love, compassion and growth. The reader embarks on a magnificent ride; one full of reflection, a strong support system and above all, mindfulness. I felt inspired after reading these essays. They equipped and prepared me to be part of the mindful revolution. This book should be read by government officials and policy makers. In a quiet and unassuming way, Prattis provides the tools we need to move forward in creating a positive world of community. His writing is accessible, not overpowering. We want to continue reading and be part of Prattis' world. This collection of writings is a gift to the global community.

- **Ginette D'Aoust-Castonguay** Wellness Facilitator:

While reading this rich collection of writings conveying pragmatic life lessons, I found *Our World Is Burning: My Views on Mindful Engagement* to be a profound yet comprehensive guide. It is capable of igniting a spark deep within us as well as inspiring the reader to be present and step up to the challenge.

- **Anita Rizvi,** Therapist:

Dr. Ian Prattis, with the vision of a Prophet, the heart of a Buddha and the mind of a master Story Teller, offers a timely and 'failsafe' gift to humanity as our poisoned collective psyche, reflected in the deterioration of our ecosystem, is poised to burn on the pyre of global consumption. When we look deeply into *Our World is Burning* we see the frightened face of our inner child staring back, nine year old James' anguished cries merely reflecting our own. And just as 'World Leaders' continue to hatch the fabricated concept of separateness, gleefully filling in spaces with nefarious modus operandi designed to further divide brothers, paying no heed to the droughts and famines that are consuming innocent children as their mothers look on with horror, continuing to stomp on Mother Earth's heart leaving ego sized footprints on every corner, relentlessly banging on war drums, stepping up the game of iniquity with innovative methods for mass destruction such as biological warfare. In the midst of it all... a Teacher has come... Now, pick up a copy of *Our World is Burning* and watch evil leave the room.

- **Melissa Studdard,** author and poet:

Amidst the fear, greed, and pain of our burning world, there is a cool garden where you can recover hope for posterity and cultivate your best life. Ian Prattis' words are one of the surest pathways to that garden. Both analytically rigorous and fearlessly honest, this book is a must read for anyone asking, "What can I do?"

- **Wendy Martin,** PhD:

An antidote to "sedentary agitation." These writings, dharma talks, and stories offer compelling examples of how to respond to the most serious social, economic, environmental and personal challenges of the Twenty-first century: inequality, climate change, addiction, consumerism, depression, violence, abuse. Through integrating personal narrative with insights from Buddhism, anthropology, psychology, and ecology, Dr. Prattis examines the interdependent nature of the self, society and nature, as well as, the integral relationship between self-transformation and collective healing. Drawing on the threefold training of concentration, insight and moral action, he illustrates how individuals can use mindfulness practice to cultivate awareness, as an ethical framework to guide actions, to create steadiness and equanimity, and to replenish body, mind and spirit. For those that may be asking, what to do in the face of … anxiety-, uncertainty-, fear-, despair-, injustice-, powerlessness-, frustration, apathy-,this book offers effective tools and strategies to help identify and transform the causes, conditions and manifestations of our individual and collective fear, suffering and anger into compassion, courage, healing, awareness and mindful action.

TABLE OF CONTENTS

"The theme of these views-essays is change, cycles of transformation and discovering how we contain everything within ourselves."

- Dr. Ian Prattis

Invitation

As an idealistic teenager, I wanted to save the world. I still do. Over the years though, I discovered I first had to save myself, because I was every bit as screwed up as the world.

Indeed, saving myself and saving the world seems to be the same struggle, because we are all connected, one to another, and the forces that warped me are the same that warp the world.

These views-essays form the chapters in this book and come out of my long struggle. Please accept them as a gift; my thoughts on how to transform ourselves and our world.

The sixteen views-essays are not candidates for academic bickering or pawns in the intellectual constructions of clever talk.

When a breeze caresses a falling leaf, it is transformed in its descent to earth. Sunlight catches one side then glances off the other as the leaf gently spirals down. The impermanence of this gift of nature is part of what makes it beautiful.

Yet, notions of permanence reflect our fear of the unknown and foster the limitations we impose on reality.

Impermanence connotes our true nature of interconnectedness with a constantly changing web of life. We are fully alive in our connection to everything else.

The theme of these views-essays is change, cycles of transformation and discovering how we contain everything within ourselves. They rest on the ever-changing cycles that mark our journey in these tumultuous and dangerous times.

The opening piece – *Our World Is Burning* sets the theme for this book and it focuses heavily on climate change and Mankind's devastating role in this major issue

Rant From the Future and *Chronicles of Awakening* draw their inspiration from and are based upon my 2016 book *New Planet, New World*.

Ottawa Independent Writers brought out a unique anthology in 2016. My part in that stellar release is *Dawson's Desert Legacy,* and I share those views in an expanded manner via Chapter 14's viewpoint-essay.

Chapter-essay 4 *Punk Palace* finds its inspiration via an earlier article published in a different form in The Shambhala Sun (September 2005). I expand upon it in this collection of writings on a wider stage.

Excellent editing by Meghan Negrijn and Michael B. Davie ensured that my essays wove an elegant tapestry about how to manifest mindfulness in our difficult times.

PART ONE:
CASCADE

> *"What was missing from the deliberations and press releases was a candid recognition of the "Cascade Effect," a notion from ecological science. Tipping points in sea level rise and temperature connect to tipping points in air pollution, which connect to tipping points in polar ice melt and boreal forest wildfires. All these triggers amplify tipping points that create deforestation, desertification and so on in a relentless cascade."*
>
> **- Dr. Ian Prattis**

1: Our World Is Burning

Leonardo DiCaprio has presented passionate videos that climate change is a fact. He draws on a unanimous scientific consensus. Not so in the Trump presidency, where climate change in the United States is swiftly being displaced.

Trump has dubbed climate change a hoax created by the Chinese government to make US manufacturing non-competitive. He tapped Myron Ebell, America's most prominent climate change skeptic, to oversee the Environment Protection Agency (EPA) with a view to roll back the extensive environmental platform created by the Obama administration.

Myron Ebell is not a scientist and does not believe in the facts endorsed by climatologists. He talks glibly about the actual benefits of climate change and rightly earned the "climate criminal" tag from Greenpeace.

Trump then selected Scott Pruit to run the EPA. Pruit is an ally of the fossil fuel industry and his selection will destroy the US Clean Power Plan and all the other environmental measures put in place over the past eight years. He proposes to open up federal lands for logging and carbon extraction – oil, gas, coal – and rejects the Paris climate change accord. Conservation is not part of his vocabulary.

The XL pipeline has been approved and federal parks will end up drastically diminished. In addition, off-shore drilling permits will be abundant and conservation measures could be dropped world-wide.

The strategic momentum engineered by these two climate change deniers makes the United States a rogue nation. But, as a powerful country, its impact will destabilize global efforts to reign in climate change. Myron Ebell's organization - Competitive Enterprise Institute (CEI) - is financed by Exxon and the coal industry. It is part of the powerful international misinformation machine that pours millions of dollars into the campaign that discredits climate scientists.

CEI masquerades as a think tank but is in fact a corporate lobbying group that buys the politics to protect the interests of billionaires, many of whom have no concern for a sustainable environment.

Fossil fuel interests greeted Trump's strategy with elation in anticipation of a new bottom line - protection of carbon profits for Trump's corporate friends. The existence of the EPA is endangered and could be cast aside. Trump and his followers give little hope for our deteriorating planet.

Though there are rumblings in Washington of impeaching or charging Trump with treason, the dice has already been rolled. His cabinet and any replacement are all cut from the same cloth. The recipe is in place for disastrous global consequences.

My prior book *New Planet, New World,* published by Manor House, is set in 2080. It charts the space mission to inhabit a new planet made necessary by willful ignorance about climate change. A culture crash late in the twenty first century opens this epic novel. Children travel via spacecraft to a distant planet to escape Earth. A sharing of cultures and technologies ensues as they join other Earth refugees to form a sustainable ethical and moral community.

Intertwining plotlines arc into the epiphany of the final chapter, a philosophy for the future. The inclusiveness of science combines with Tolstoy's vision, Pope Francis' climate change encyclical and not repeating the mistakes of the carbon cabal.

The underlying message is from Tolstoy, who is considered to be the 'Conscience of Humanity.' He described humanity's bottom line as the cultivation of love, the mainspring for authentic and responsible living.

The final chapter – Musings on the Future of Humanity – was written long before Trump ascended to the presidency in America. However, readers pointed out that I provide an antidote for all that Trump intends to implement.

I now bring a sensitive perspective to your attention, by seeing climate change through the eyes of a terrified nine-year-old boy. My grand-nephew James was recently celebrating his birthday, yet he felt awful about being nine. He wished he could stay five years old forever. When asked why, he replied that if he could stay five then the Earth would not explode. His lips quivered and tears welled up in his large brown eyes. He said, "I don't want to grow up and live in a world that is burning."

In the silence that stretched between us I wondered what to say. I could not say that everything will be OK. He was much too intelligent for such placebos. So I spoke to him about the mindfulness community I created in 1997 – Pine Gate – and the deliberate steps taken for planetary care. We simplify, make do with less, share and adapt.

Our intent is to create environmental leaders and that includes him. "Why not become a leader for your generation?" I asked him. He thought about that and asked what else did Pine Gate do?

I pointed out that Pine Gate encourages voluntary simplicity and community ethics as a way of life. We start with the earth. Our organic garden produces an abundance of vegetables, apples and flowers that are shared with neighbors and community members. It is a solace for me to spend time with the Earth, observing bumblebees and butterflies while gardening with assistance from neighborhood children. I told James that the kids once laughed hilariously when they saw that the vegetable plant I had carefully nurtured turned out to be a giant weed and not a tomato plant!

We had great fun returning it to the compost bin. At the back of the garden is a beautiful fountain that murmurs next to the flowers, which are sent to the elderly folk living on our crescent. A simple underground economy arises from the sharing.

A solar panel on the roof fuels the hot water system of our home. Everything else is as eco-friendly as we can make it for a 50-year-old bungalow with a meditation hall in the basement. This eco-effort has become an example for other friends as they consider how much we are saving and implement something similar.

Our focus is on mindfulness in schools and city environment, teens at risk and the empowerment of women. I admitted to James that I am shocked by the results. At the local level there were great women who helped make things happen.

"You mean girl power?" asked James in surprise.

"Exactly that," I replied, and told him about my writing that the present millennium is the century of daughters, not so much as gender separation, but as attributes of a holistic, nurturing presence of mind.

The idea behind Pine Gate is to foster a strong group of people in Ottawa making a difference for the betterment of society and the Earth. Women are in the forefront of this endeavor. They are the heart that holds the living waters, the dynamic epicentre that leads to effective action. That is how we get things done, creating a different course of action and living.

James was taking it all in, instinctively knowing that major changes were needed. I suggested that when enough of us change, then our ideas will be in charge. I told him about a speech I had given about violent consumption and pointed out that festive occasions like Christmas provide opportunities for the best and the worst within us to come out and play. Yet compassion and kindness are quickly overshadowed by greed, selfishness and consumer madness. We need to move on from being self-absorbed and distracted.

"How?" he asked again, as he really wanted to know. I chose my words carefully.

"We need to locate in something bigger than ourselves; a humanitarian cause, respecting the earth, while making our thinking better, being kinder and more generous.

How about examining our habits about gift giving and learn to give gifts that make a difference? I no longer buy Christmas gifts, instead present gift certificates in the name of family, grand-children and young neighborhood friends. These gift certificates include items like education for a girl in Afghanistan, micro-loans for female led families, rebuilding forests in Haiti, literacy packages and mosquito nets where needed, support for Habitat for Humanity building houses for the destitute and so on. Such gifts are bigger than us and create happiness for less fortunate people."

I told James how my grandchildren proudly take their Christmas certificates to school for Show-and-Tell periods. They play it forward with their class mates and teachers. One boy on the crescent where I live has received such gifts for several years. For his most recent birthday he asked all his friends not to give presents, but to bring a donation for the Ottawa Humane Society that looks after hurt animals. All of his friends brought donations, a splendid sum of one hundred and eighty dollars. They all went together to the Humane Society and happily handed their bag of cash to the surprised staff. Other children in the neighborhood have followed suit.

This resonated with James. He said, "I could do that with my ice hockey team. My dad is the coach and he would help." He waited for me to continue.

"James, the greatest gift we can give to ourselves and others at this time of global crises is sharing and caring. It involves stepping onto what the Buddhists call the Bodhisattva Path." (James knows that I am a Zen teacher).

I explained that a Bodhisattva was a person who stayed in the global mess and did their best to awaken the minds and hearts of people. I firmly stated that it is time for the Bodhisattva-within-us to enter the 21st century as the example for action. It takes training, practice, intelligence and creative vision.

"You mean like Jedi training?" he enquired. I nodded with a smile. I referred briefly to my years of training in ashrams and monasteries in India and France and with indigenous medicine people. I confided that the real kicker for me was the time spent alone in the Canadian wilderness. I promised to talk to him about this later.

"So what is the big deal about violent consumption?" he asked.

I replied that it totally dominates our planet, mind and body. I tried to explain how, knowing that James' greatest fear was about the planet's ecological crises. He worried about mining disasters in Brazil and China, wildfires in Canada's Boreal forests, Amazonian deforestation and the Gulf Oil Spill where tons of toxic oil dispersants contaminate the oceanic ecosystem.

"How do we change the destruction of the planet?" James exclaimed.

I wondered how best to explain matters to him, yet trusted his intelligence. I said, "We must come to a stop, locate ourselves in stillness and make different choices by examining our minds and consumption patterns. We must look at how we actually participate in creating these terrible disasters." I noted that this kind of awareness takes us back to what we do with our minds.

"Just how?" was his one line mantra.

"You can start by making friends with your breath," I said. James looked up at me quizzically.

"You just bring your focus and attention to your in-breath, then on your out-breath. Really concentrate on the whole length of breath in and breath out. Do this ten times. This kind of focus peels away anxiety, frustration and anger so that you become calm and clear. Try it with me and notice the difference for yourself."

He did so, nodded and grinned with agreement. I told James that we do know how to reduce our ecological footprint. We also know that taking care of the earth and the oceans takes care of ourselves. We must begin it now for the future, as our tomorrow is shaped by the actions we take at this moment. I looked at James and suggested that was enough for him to digest, but he yelled, "No, I want to hear more."

I could not turn away from his eagerness. I mentioned that if rampant consumption remains our deepest desire we will continue to degrade the planet, eventually destroying its ability to harbor life.

His fears were correct. Valentine's Day, Easter, Christmas, Mother's Day and so on are targeted by the captains of industry for optimal retail returns, and mindless consumerism is fuelled to the max. At Christmas we are far removed from remembering the significance of this spiritual celebration. It feels like endless economic growth, the mantra of modern civilization, pushes expectations with no awareness of the consequences for our own health or the health of the planet.

Our current non-sustainable energy and economic systems are subsystems of a global ecology that is disintegrating before our very eyes. We must simplify, make do with less and change, or the burning world will definitely occur.

"Did you know that we also harm our bodies through the food we eat, and that it has disastrous consequences for our connection to all living beings?" He did not, yet his mind was a sponge soaking up every word. I continued, "The vast consumption of meat and alcohol creates an excessive ecological footprint. Industrial animal agriculture is not really farming. Animals are treated solely as economic commodities and subjected to horrible cruelty. The stress, despair and anger generated in the animals are the energies we consume when they end up on our plate."

"That is so gross," remarked James.

I told him that we can change our minds and patterns of food consumption. We re-educate and retrain ourselves mentally, choosing to support our body and planet by shifting ingrained food habits. It takes training but we can begin to step more lightly on the planet.

It means reducing as much as possible the violence, destruction and suffering brought to living creatures and to the planet. Bringing peace into our own biological system and consciousness inevitably brings it to all the other systems that we engage with through our thoughts, speech and actions.

"Is this your Buddhism?" James then asked.

I smiled, "The Buddha was very smart. He taught that the world is always burning, but burning with the fires of greed, anger and foolishness. His advice was simple; drop such dangers as soon as possible. What the Buddha taught was that it was the unskillful speech, selfish feelings, negative mental formations, wrong perceptions and badass consciousness that burned the world.

James laughed, "Did the Buddha really use the term badass?"

I grinned and said that was my embellishment, then pointed out that the Hopi people also referred to the burning as a state of imbalance known as *Koyaanisqatsi*.

We are not the first people to experience this. The difference today is that without our commitment to wise intervention, we could be the last.

"Is climate change our basic problem then?" he asked.

I paused for a moment before replying. "The basic issue is whether we can adapt to climate change. You know about the 2015 Paris Accord on Climate Change. We talked about it before." James nodded. "It was an exceptional step by the international community, showing their determination to prevent global temperatures from rising a further 1.5 degrees. The signatories returned to their respective countries to find the wherewithal to - "Change Climate Change."

What was missing from all the deliberations and press releases was a candid recognition of the "Cascade Effect," a notion from ecological science. Tipping points in sea level rise and temperature connect to tipping points in air pollution, which connect to tipping points in polar ice melt and boreal forest wildfires. All these triggers amplify tipping points that create deforestation, desertification and so on in a relentless cascade."

I reminded him of the wildfires in Alberta. "It was not a singular disaster at Fort McMurray, as the entire boreal forest in Canada is a tinder box due to climate change. The reality is not about a reversal but about learning how to adapt to the ***consequences*** of climate change." I emphasized to James that the disasters all over the world interconnect. Whether wildfires, floods, landslides, volcanic eruptions, hurricanes, tsunamis or millions of aquatic creatures dead on beaches, it is all connected. The media and news reporters cast science to the wind when they report the drama and hype of terrible things happening world-wide but rarely tell the truth that it is another manifestation of climate change.

News programs are often focused on ratings and some openly promote corporate interests that are contributing to these interconnected disasters. The general public is not educated by the media about the terrible realities happening on our planet.

Other obstacles that prevent the general public from taking wise action are a mixture of fear, despair, laziness, disempowerment and a sense of hopelessness.

"What on earth can I do to make a difference?" is a phrase muttered all over the world in countless languages. Followed by, "Why should I do anything?" There is certainly global awareness, but also fear about our future place on this planet. Maybe this is why you want to stay five years old forever. The difficult thing for you, for anyone, to grasp is that we are the primary cause."

I confessed to James that in my previous books, I had underestimated the impact of the carbon fuel cabal, a complex web of powerful corporate and government interests. "This carbon economy extends into the manufacturing and servicing sectors, supported by financial institutions that control marketing and advertising. This collective power, when extended into the media, has attempted to make science and ecology into public enemy number one. The result is a push to circumvent the Climate Change accords agreed to by the international community. People everywhere are aware, but just feel helpless in the face of this power."

So what are we supposed to do?" James shrugged in exasperation.

"Here's the thing," I said. "In terms of action, we have clear data-based evidence that we must cut back, make-do with less and implement a lifestyle of voluntary simplicity. So, where do we start? Of course we must think globally and be aware of the bigger picture despite fear and disempowerment. But we can also act locally in our families and communities. Our intentions then spread like ripples from a pebble dropped in still water. We can hold officials, politicians and corporate culture to account. We can tell the politicians and corporate decision makers that we, as voters and consumers, are deeply concerned about the planet and our impact on it."

I continued speaking on a personal note, "So James, the challenge for me is to be *in* society, but as a *still island of mindfulness*. Take small steps at first, then larger ones. We just need to make essential changes in energy use, diet, language, media and outreach. Voluntary Simplicity is a good starting place. It means making deliberate choices about how we spend time and money. We can support environmental causes with excess clutter in the basement and always think about whether we really "need" to buy something more. Enjoy being simple and living modestly by shifting our perceptions a little bit. Looking deeply into what we do with time, money, clutter and our choices, *we can change*. Then see whether the consequences are peace and happiness for you. The world will follow."

I told him about my futuristic book – *New Planet, New World* - which provides a counterpoint story to the demise of our modern civilization. In this book I chart a communal Hero's Journey to reconstruct society based on ecology, caring and sharing. The final chapter muses about human survival anywhere. The drive is to create a tangible spirit of co-operation, the willingness to share and be supportive and teaching how to cross the bridges of misunderstanding. My main intention was to provide a reflection of the disasters of the world today. In the novel the rich and uber-wealthy already inhabit armed, gated communities and will be targets for eco-militias and popular uprisings drawn from the impoverished masses intent on revenge.

"Have you seen Stanley Kubrick's *The Clockwork Orange?* film" James had not. I told him it was a gruesome movie that could well emerge in the real world. "To avoid this, it is wise to take training very, very seriously. This helps with all the negativity I was telling you about"

"Wow," exclaimed James. "OK, I get it about training but what does it look like?" I was relieved by his intelligent questions but hesitant to talk to him about what I was thinking. Then he said: "Just lay it out for me."

I then proceeded to talk about "Gardening in the Mind." I offered him eight simple steps to refine the mind and then engage differently with the world.

1. You – learn to be silent and quiet. Clear time and space for spiritual practice at home and throughout your daily schedule.

2. Create a stress reduction menu and subtract the "weeds" in the garden of your mind.

3. Be determined to meditate daily – do the weeding.

4. Focus on and soften your heart – nurture the soil of your mind's garden.

5. Cultivate the seeds of mindfulness at home, school, and work or in solitude.

6. Simplify, make do with less, de-clutter your mind and home.

7. Taste the fruits of your spiritual practice that change your mind.

8. Engage with the world.

James was typing all of this down on his tablet as I continued to talk. "Our ways of living together, caring for environmental, political and economic realms need to be re-constructed." I assured James that we have the capacity to transform the mind. Finding stillness and inner silence is a necessary first step. We have to find a way to create the conditions for this to happen. In our modern world of fast paced lifestyles there are so many distractions that make us outwardly dependant and un-centered.

We also find it easier to close down rather than open up our hearts. But the remedy is within reach. We can unravel the knots of suffering and move from being mindless to being mindful, achieved by gardening in the mind."

I paused for a while to find the words to bring our conversation to an end. "Why should we do this stuff James? Here's why. When you are open and receptive you become an epi-center of light and energy for others.

When you can sit with pain, face to face with what hurts, breathing in and out, you feel the sting recede as you calm. If you start to close down ask yourself, "Do I really want to take a pass on happiness?" Always let go once you feel you are closing down or clinging."

Then I said to him, "Do you know that I have a fridge magnet at home with the words - LET GO OR BE DRAGGED? I see it every day and take the message to heart with a quiet smile. It is essential to learn to be silent, to stop clinging and find the way to be present in the moment. As the Hopi advise us, never take anything personally and look around to see who is with you. Doing these things helps the world change. Such a destination is well worth your effort don't you think?"

James nodded his agreement.

My conversation with young James was all about the essential teachings of the Buddha. Engaged Buddhism is a modern term coined by Thich Nhat Hanh to remind Buddhists that the Buddha's teachings were always based on Engaged Buddhism. In the past there was too much attention on forging feudal structures to support monasteries in the East and the foundation of Engaged Buddhism was lost. It is up to us to revive it and live it in every moment of our lives. If the reader connects the dots of my conversation with young James, you would see clearly that Engaged Buddhism is one antidote to all that Donald Trump's government stands for.

I assured James that we are equal to the task and I chose not to hold back anything from him during this long conversation on his birthday. He is an unusually bright boy, asked questions and demanded clarification. Yet I knew he had grasped what I had said. He came up to me as I was leaving and whispered in my ear that my chat with him was his best birthday present ever.

2: **Rant from the Future**

(To convey the serious implications of uncorrected climate change and environmental damage, I have created a fictional yet all-too-possible future scenario in which a conscientious scientist presents the speech of a lifetime before an audience of self-serving power-brokers who have allowed corporations to inflict lasting environmental damage on Earth, beyond the tipping point, to the extent that our planet now faces the extinction of human life. This is unfortunately a depiction of a potential and indeed likely outcome if we continue on our destruction path.)

YEAR: 2080   LOCATION: UN   CRISIS: Earth Imploding

Dr. Tom Hagen's blistering speech to an elite forum of political and corporate leaders at the United Nations in Geneva, Switzerland in 2080 changed the future of humanity.

Gathered before him were power brokers from around the world.

Tom was an astrophysicist, engineer and prolific author of searing plays about human fragility and books for children to inspire them to care for the Earth.

His endeavors, however, did not turn the tide of ignorance about climate change despite his creation of detailed scenarios for adapting to climate change and reforming business ethics.

Tom was also chef-de-mission of the International Space Agency's PRIME 3 project to locate a suitable planet, as habitation on Earth became increasingly compromised. The project was outlined in Space Agency folders that each member of the audience had before them.

The International Space Agency had established research stations that orbited Mars and the agency had also established stations on the surface of Jupiter's smallest moon, Europa.

Europa had an iron ore and rocky mantle, similar to Earth, yet the rocky exterior was covered by a 100-kilometre-thick layer of ice. An ocean was identified beneath Europa's frozen crust. Radical advances in space technology made this possible mid-century, through the invention of space elevator nanotubes and revolutionary advances in nuclear fusion engines for spaceships.

Station One at Jupiter was a key construction. From there a probe, PRIME 1, was launched into the heliosphere through a wormhole into interstellar space. In a neighboring galaxy it located a planet with two moons in an ecliptic plane with a dozen planets orbiting a massive sequence star, the sun for this system. The planet had a liquid hydrosphere similar to Earth.

A more sophisticated probe, PRIME 2, sent back information identifying distinct zones from tropics to polar with evidence of oceans, forests and mountains. No sign of a sentient life-form was revealed.

Both probes identified, in the upper stratosphere of the planet, a dense particle field, similar to the Van Allen belt.

The long-term plan was for Jupiter One to serve as a way station to ferry pioneers to the new planet.

The Jupiter One project required massive financial support that Tom was trying to elicit.

Tom stood quietly at the podium in Geneva, about to speak.

He had both good and bad news for his powerful and wealthy audience, deeply hoping some of them would finance this late opportunity for survival of the human species.

Tom looked at his carefully researched notes and put them to one side. This speech, the most significant of his life, had to come directly from his heart.

He composed himself, standing still and silent at the podium, six foot three inches of intense focus, dignified and alert. Taking his glasses off, he placed them on top of his notes and glanced at Sian, his wife, sitting off to one side of the podium. She smiled with relief as she knew what was to come. His fifteen-year-old daughter, Catriona, was sitting with Sian in the front row, her smile beaming. She wore a fashionable grey pant suit that offset her lustrous red hair tied up in a bun. Her mother had fetched her from her boarding school in Switzerland, knowing that Catriona would love to hear her father speak.

Tom looked around at the audience, one he did not particularly like, but had to convince. He breathed deeply and waited for calm, feeling that icy steel of reason and vision to ignite his insight.

He began to speak in a calm, clear baritone voice:

"In this moment what is left of the population in Australia is being evacuated. The interior sand storms and volcanic eruptions in eastern Australia plus successive coastal tsunamis have brought an end to human occupation there."

He paused a moment to let what he had just said sink into the consciousness of the audience, then continued:

"The inundation of Bangladesh, the Netherlands and coastal regions around the world are a direct consequence of the collapse of the polar ice sheets, which increased sea levels by seven meters. These events have dispersed half the global population, ushering in plagues and pestilence that have eliminated ninety per cent of animal species and directly threaten human survival. These are facts that cannot be refuted. Furthermore we cannot turn away from the reality that our entire planet is desolated and overwhelmed with refugee camps."

Tom paused for several moments, making eye contact with individuals who could support him. He glanced towards Catriona, still smiling her support and it helped him gather his wits. The facts he had just delivered set the tone for what was to come. He knew many of his listeners would be offended but it did not stop him.

"In the early part of the twenty-first century the leap to a zero-net-carbon world was possible to make. Yet the opposite trajectory was chosen with a rapid increase in greenhouse gases. Instead, wealthy nations and economic enterprises doubled their production of fossil fuels."

He raised his voice, "Did you not notice that degradation of the Earth's ecology was the catalyst for radical climate change? Did you not see that food crops were destroyed by horrendous heat waves? Did you not realize that food riots and world panic could be traced back to the economic agenda of energy extraction? Collective power was invested in political, social and economic structures that centered on the carbon combustion complex and this collective agenda destabilized world order."

Several members of the industrial elite stood up and left. There was another pregnant pause, as Tom waited for them to walk out without comment.

"This focus on economic wealth at all costs was short-sighted. The global financial system ignored the welfare of populations and the ecological breakdowns caused by these actions. May I remind you that the economy is a mere sub-set of the mother lode of ecology and most ecosystems on Earth have been destroyed. Financial collapses signaled dangerous global watersheds. The world food system crashed from the chaos brought in by climate change. No one asked different questions or tried to find different answers. The anger of the populace then turned on the powerful masters of capital and politics. At the extreme end of the violence spectrum, anger boiled over into lynching corporate and political leaders held responsible by eco-militias. The heinous actions of these black clad anarchists are certainly an extreme response to the control, lies and doublespeak."

The inherent danger registered with the rich and powerful still present. Tom continued to speak, "Look back over this century and see why such violence emerged. Millions of people have died from thirst, starvation and disease. Death arrived from every pestilence, some of it created in your counter-intelligence labs. Those countless millions who have died do not include those lost in the many wars waged over scarce resources, particularly water. Those wars have their origins in your greed for money, control and power leading directly to the cascade of disintegrating eco-systems. Government, industry, banks and financiers grew wealthy by ignoring the ecological consequences. I have spoken before to gatherings of this nature and provided a warning to all about the course we are on. I will try one more time."

Tom felt exasperation arising within him yet calmly continued along the same track:

"Your dismissal of dire warnings served to discredit climate change scientists and oceanographers who proved that eco-systems were disintegrating. You silenced and jailed citizens with the integrity to save the Earth. But it was never about the unanimity of science or free speech. It was about the brand of economics favored by your fossil fuel complex, a collective cabal of extraordinary power that extended its reach to encompass all powerful corporate ventures."

Tom paused to sternly stare at his audience and in an angry tone, he stated: "You single mindedly created a powerful culture of denial about climate change and how it has impacted the cryosphere to such deadly effect."

"We all saw that social order broke down in mid-century ushering in the overthrow of governments, the establishment of martial law and Nazism. All of which increased the desperation of populations worldwide, who took to the streets in mass riots. In the vacuum of social order, opportunistic, vicious warlords and militias took over much of the world. You've all suffered from violence of so-called eco-militias, which hunt down and string up politicians and corporate leaders they hold responsible for the collapse of Earth's eco-systems. Let me be clear. These eco-militias are pathological criminals, yet they constitute an ever present danger to our entire planet. You and I have lost many friends and colleagues to the murders they have carried out. They are extreme fanatics, but you are equally so – fiddling with indifference while Rome burned. The entire planet has been allowed to burn on your watch. Perhaps even at this late stage you can learn something from Rumi, the Sufi saint. He said, 'Sit down and be quiet. You are drunk and this is the edge of the roof.'"

Tom allowed an entire minute for that to sink in before continuing:

"Your policies and brand of economics have forced humanity off the edge of the roof and you can now see the consequences worldwide. Oil consortiums and government created incredible propaganda campaigns to promote oil and gas extraction, irrespective of the damage caused to ecosystems and populations. They produced false images of reforestation, utmost safety, deep concern for wildlife, populations and clean water. This played to receptive audiences yet decades later we find rivers and lakes occupying a wasteland."

Tom paused a moment, and then continued: "Oil derivatives swiftly poured through interconnected waterways. Aboriginal populations world-wide that once augmented their households with fish, game and forest products are no more. They relocated or died. This contributed to the impossibility of any form of transition to a sustainable, renewable economy."

His words secured the exit of more power holders, though his next speech caused a few of them to stop in their tracks:

"The billionaires amongst you have well equipped and tightly defended underground bunkers to escape to, but I have bad news for you: The deep core drilling for oil and gas all over the world, particularly in fragile ocean beds, has compromised the tectonic plates at the center of the planet. This will cause world-wide volcanic eruptions and earthquakes, destroying your underground bunkers. It is estimated that millions will die from the volcanic explosions, leeching poisonous gases and the inevitable tsunamis in every ocean," Tom said, reminding his audience that there had been many early warning signals:

"Seventy years ago seismologists provided the critical evidence that deep core drilling and fracking were directly associated with the rapid increase in global earthquake and volcanic epicenters. The collision of tectonic plates is now triggering massive earthquakes and volcanic eruptions. Repeated warnings from scientists to ban deep core drilling were ignored, bringing the unspeakable into reality."

Tom paused and took a deep breath before looking around at his depleted audience.

"I ask you at this late stage to do one thing. Take a good look at the dossier provided by the International Space Agency," he said.

"Support the PRIME 3 space project to the tune of fifteen-billion dollars. That is what it will cost to build a new spaceship and create a viable outpost at Jupiter Station One. The very future of our species lies in your hands. PRIME 3 is the last opportunity to begin anew, to avoid replicating the structures and policies that have led to the inevitable demise of Planet Earth," he added.

Tom stopped talking, greeted by a cold silence. There was no applause, no acknowledgement.

He walked steadily to the side of the stage where Sian was standing, a look of admiration on her face.

Embracing him lightly, she took his hand in hers as they left the podium.

Catriona stood up, taking her father's arm and joined her parents as they walked up the center of the UN forum. She was so proud of her father, despite the audience's unwillingness to meet his eye. Sian sent a smile to everyone, regardless of their outlook as they passed through the security of the UN building.

Walking down the graceful stone steps, Tom let out a long breath, as he glanced at Lac Geneva sparkling in the distance.

"I sure blew that one. Hardly any one of those people allowed what I said to sink into their mind. I have no idea where to turn to ensure the PRIME 3 project gets funded."

"Perhaps you may be wrong Tom," Sian replied. "Look over there."

An armored limousine had drawn up in front of them and heavily armed guards quickly fanned out in a defensive format.

One guard opened the side door of the armored vehicle.

Out stepped Seymour Hansen, the president of the biggest bank in the United States. A tall, commanding figure, his greying hair was offset by deep flinty blue eyes fixed upon Tom.

"Dr. Hagen, you do know who I am?"

Tom nodded.

"I listened to your powerful speech. You may be surprised to learn that it impacted me deeply. I have sent a line of credit to the International Space Agency for the fifteen billion dollars required for your project. I am not buying a seat for myself on your space craft, as I doubt that I possess the necessary credentials. However, I have lined up a good consortium of technologically sophisticated corporations. They will provide whatever technology and systems you need."

Tom was stunned, yet asked, "Why are you doing this?"

Seymour Hansen replied with a wry smile, "Your speech stung everyone, some more deeply than you realize. I admired your courage as well as your vision and precision. As far as I am concerned I would not trust anyone else with this space project. I and the others may not be alive to see its fruition but there are no strings to my offer."

Tom stood dumbfounded at hearing from the most powerful oligarch in the world. He did not know what to say.

Thankfully, Sian stepped forward to their benefactor and gently kissed him on the cheek, simply expressing their gratitude.

Hansen smiled, his flinty eyes softening, "It's good he has you Mrs. Hagen for the necessary graces."

He shook Tom's hand, "Dr. Hagen, get it done."

Tom nodded, as his heart expanded with relief and determination. Sian was delighted that Tom's vision had been supported and his daughter giggled at the bewildered look in her father's face. Sian had to shake him by the shoulders to ensure it had sunk in.

3: Are We Stupid?

Oscar nominee Pete Postlethwaite plays the best role of his acting career in the film *The Age of Stupid.*

The movie fast forwards us to the year 2055. Pete plays the sole character in this riveting film. He stars as an old man living alone in a world totally decimated by climate change. His location is the High Arctic.

The film makes James Lovelock's conclusions in his 2006 book *Revenge of Gaia,* seem prophetic. In an interview about this book, Lovelock provides a dire prediction for humanity: "Before this century is over, billions of us will die, and the few breeding pairs of people that survive will be in the Arctic where the climate remains tolerable."

The film character played by Postlethwaite is the curator of The Global Archive, a digital storage laboratory located in the Arctic. It is the last habitable place for humans on Planet Earth. The footage he views shows how climate change reached tipping points and runaway effects while at the same time humanity's achievements are also saved for posterity.

He wonders, how could human minds capable of such monumental achievements neglect the destruction happening to their lived-in-ecosystem? The old man shakes his head in disbelief while looking for an answer. The film gives us an answer – carbon based energy. Our addictive dependence to it is what propelled the downward spiral of devastation. The addictive process was enabled because we allowed the environment to become an extension of human egocentric needs and values, an ego-sphere rather than an eco-sphere.

In this ego-sphere, we consumed mindlessly in the global economy without regard for ecosystem balance or concern about creating inequality, poverty and ecosystem imbalance. Planetary care is not on this agenda, as the film graphically shows. We see the old man in the High Arctic watching archival video footage, carefully preserved from 2008. His stark question to the viewer is: "Why didn't we stop climate change in 2008 when there was a chance?"

The director of the film, Fanny Armstrong, creates a montage from live news and documentaries saved from 1950 to 2008. The video record charts the steps taken by humanity into global devastation – devastating that is for human habitation and for all other species.

In an artfully created mosaic in this film, six real life characters play out the dramas of their personal stories. Their humanity and incredible stupidity are extant in this brilliant tapestry of human folly. What is so gripping is that we who view it are made to feel distinctly uncomfortable, because their shadows and myopia reflect our own. They arrive as a projection of our political and corporate leaders.

After watching this film, we can no longer hide from these shadows. We are forcibly held to account. If we do not act now, this film then becomes our story.

Two Forks in the Road – Which One Will We Take?

'Failsafe' is an engineering term used to describe a lever or stop valve that comes into play when a piece of machinery is just about to self-destruct. Phut!

The lever comes down or the stop valve kicks in before the boiler blows up or the nuclear core melts down before inevitable destruction occurs.

I talked about the Failsafe in Consciousness concept in my book, *Failsafe: Saving the Earth from Ourselves,* published by Manor House in 2008. It describes how consciousness expansion will be held back by a deliberately cultivated ignorance about better knowledge. That is, until the global ecological situation deteriorates to a breaking point. My thought was that this breaking point will then act as a catalyst, exposing such ignorance. At which point consciousness would be propelled into expansion, deliberation and change.

My vision was a positive one, as I believed that humanity could create new structures and organizations. Out of these would emerge the radical solutions to address the ecological emergency we all face.

We have the knowledge to create this solution, but the obstacles that stand in the way are not technological. They are the attitudes, values and concepts that define the present dominance of corporate values, rampantly consolidated through "turbo-capitalism."

I argued that the necessary clarity to deal with the global environmental crises will emerge, once our thoughts, values and attitudes change and no longer sustain and feed our internal pollution. This is the radical internal climate change necessary to engage intelligently with the external climate change.

There is certainly global awareness, but also fear, about the future of Earth. The overwhelming terror of Gaia crashing down on us is unbearable.

Many years ago in India I had an audience with Sai Baba. I was visiting this sage's ashram in Andra Pradesh with an Indian friend. As he slowly walked through the morning gathering, to my utter surprise Sai Baba stopped in front of me. He spoke to me for quite a while. Somehow he knew of my commitment to environmental concerns.

I remember very little of what was said, except for one sentence that blazed into my mind and stayed there. Sai Baba said to me that a transformation in human consciousness required two per cent of the population to meditate on a daily basis. I have no clue about the knowledge source for his pronouncement, but I do remember vividly the "buzz" of energy in my mind and body when I heard it.

I translated this wisdom into a two per cent option. If only I, and others, could encourage two per cent of the people we knew to change their lifestyles to one of voluntary simplicity then the environmental crisis could be mitigated. If everybody did so, then the planet would remain habitable for all species. This would involve conserving energy usage, being aware of the effects of mindless consumerism and completing one eco-friendly action every day.

This may seem naïve, but to me the two per cent option was readily do-able and within everyone's grasp. The end result of a transformed consciousness would lead to different questions being asked, with different solutions and structures created. There would be a new mind-set to make the necessary decisions for change. This one statement from Sai Baba changed my thoughts about awakening. Not everyone has to "wake up" – just two per cent. This spearhead would provide a catalyst, the strategic tipping point, for an immediate change in planetary care.

To make Failsafe in Consciousness a robust concept I identified three interconnected components:

1. Innate Earth Wisdom,

2. Counter Culture

3. Tipping Points in Consciousness.

We do in fact possess innate earth wisdom. 99 per cent of our evolution as a species relied on hunting and gathering. This adaptation known as foraging is a strategy based on sophisticated ecosystem knowledge, integrated into harvesting patterns through a spiritual understanding of the world. That is still hardwired into our brain and I thought it was simply a matter of remembering what we already possess. My anthropological logic pointed to the retrieval of this mindset in order to activate the feedback cycle needed to prevent further degradation of the global ecosystem.

The modern-day counter culture pulled together the Ecology of Ideas from Gregory Bateson, Rachel Carson's Ecology and the fostering of Gaia as a social movement through feminists, environmentalists, and the New Age beads and incense set. This unusual coalition established a broad consensus, providing a foundation for the new science of Eco-psychology. Bateson, a prominent 20th century thinker focused on the understanding of cybernetics and ecosystems, demonstrated how our modern context has rules that need changing. He showed how ecology is a set of interconnecting feedback loops that include everything.

When we destroy some of the interconnecting loops, an ecology of ideas is created that reinforces other bad ideas. Bad, that is, for the health of the ecosystem and its components.

For the corporate world, Paul Hawken's 1993 book *The Ecology of Commerce* led the charge of re-evaluating commerce and redesigning finance capital. The wisdom of natural design is built into Hawken's call for a restorative economy, which an increasing number of manufacturers are implementing. This has prompted the emergence of a genuine environmental capitalism as opposed to the corporate "green-washing" that pervaded the 1992 Rio conference.

Some corporations appear to be subtly changing despite the knowledge that shareholder interests come first. The new discourse in business uses terms such as sustainability, civic duty and corporate citizenship in annual reports and press releases. This new language form can be "gimmicky," making it difficult to distinguish between the fake and the genuine. But CSR (Corporate Social Responsibility) is here to stay. It is now a component of globalization, endorsed by successive World Economic Forums held in Davos, Switzerland.

The present "buzz" about CSR paints corporations as protectors of Mother Earth. In many cases the green-washing public relations exercise provides a totally unearned image for corporations. Yet it cannot be dismissed as mere smoke and mirrors, as there is substance here. But could these initiatives be too little, too late?

There are signs of change. Rock stars, government initiatives in Scandinavia, ground up responses from municipalities in the United States, green energy companies, climate crusaders in the corporate world, citizens' willingness to change lifestyles – all taken together indicate that the tide could be turning. But can it penetrate the corporatization of our value system and the fate of our planet?

The restructuring of capitalism requires that social capital and community sustainability become just as important as profits. There needs to be an ethical structure of profit, providing a new direction for globalization.

The present structure has caused so much destruction to the planet and its populations, so structures and mindsets must eliminate the control exercised by international finance capitalists. There will be no post-environment economy for them to exploit! The mental shift to bring this about seems to be happening worldwide.

Over two million groups, NGOs and foundations worldwide are addressing the issues of sustainability, ecology and climate change in a comprehensive manner. There is a natural mixing of social and environmental justice with peace issues, and global grassroots activism is restoring the earth's capacity to endure. At the same time this social movement refreshes our own capacities to endure and change, as we are all intricately and intimately part of Gaia's ecosystem.

An immense global response by citizens will certainly elicit an equally massive government and corporate reply, as the bottom-up movement and top-down strategies for drastic change meet and integrate. There is not room in this global ecological emergency for separating into "US' and "THEM" categories. We are totally interconnected whether we like it or not. We will all live together or we will all die together.

An intelligent green ideology embedded in everything we produce and market is a means to bridge competing agendas. Our dependence on fossil fuels could reduce because we are aware of the deadly consequences of our addiction to oil and coal. The transition to a carbon neutral global energy system over the next few decades will be costly and require massive support from government and corporate leaders to initiate the second industrial revolution. Absolutely necessary to blunt the impact of climate change, it is a global industrial project that governments and corporations can bring about, supporting citizens. Climate change has certainly entered public consciousness but has to penetrate the corridors of political and corporate power. As global citizens we must find the ways and means to support the shift in consciousness at all levels of society to make this happen. Our future existence, and the existence of other species on Earth, depends on making a new beginning for all of us.

Tipping points in consciousness are about achieving a critical mass for radical change.

Research clearly demonstrates we are not necessarily stuck with present mindsets, although it takes extensive and diligent internal work. Just as there are tipping points in the external ecology of Gaia, so must there be tipping points in the internal ecology of consciousness.

Recent studies that use sophisticated MRI scans on the brains of Buddhist monks in meditation, demonstrate that long-term meditation practice rewired the chemical and physical structure of the brain. As a consequence, the rewiring promoted attitudinal changes in the direction of balance, harmony and happiness. New neurons and synapses are generated as a consequence of meditative processes activating memory functions in the brain.

It is clear that meditation retrains the mind by changing our brain structure so that behaviorally we transform. These recent novelties of scientific collaboration provide good news. We are not necessarily stuck with the mind state that has created a devastated and discriminating world. This obvious conclusion is both encouraging and exciting.

This brings me to *The Age of Stupid* as a watershed film. You will not be the same after you have seen it. It is impossible not to be moved. I refer the reader to *Failsafe's* Appendix I: Simple Steps to Empowerment, which provides an action plan for the global ecological emergency. The steps are:

1. Take Action

2. Get Up Close And Personal

3. Reduce Your Ecological Footprint

4. Guidelines for Business and the Workplace

5. The "Big" Picture for The Future

6. Science and Diversity

7. Environmental Organizations

8. Warning to Governments

If only we can get it right and get it right now!

The hopeful trajectory is that our diligent mindful engagement will change our brain structures to permit new paradigms of behavior to come into form. As cells in the ecosystem of Gaia it is as though humanity has aligned their neuronal networks with principles of ecosystem balance, ethics and responsibility. The critical mass has arrived and it amounts to a collective tipping point for our species. Once the negative mind is reined in then clarity and compassion provide the basis for how we can exist with the planet and with one another in a totally new way. This is what could happen if we "Begin It Now,"- the concluding words to *Failsafe: Saving the Earth From Ourselves*.

The right conditions have been created by our choice to cultivate different patterns within our minds. Thus consciousness expansion can no longer be held back as a radical internal climate change has taken place.

We interconnect with a vast counter culture that, together, is no longer a minority. We become another light shining in the quiet revolution that has over two million organizations world-wide pursuing constructive change.

The Second Fork: A Failed Genetic Experiment

However, I underestimated the lure and power of the second fork. Should a failsafe in consciousness prove to be unfounded, we are then faced with the likelihood that humanity is a failed genetic experiment.

If we continue to turn our beautiful rivers into sewers, it is obvious that there is no place on Earth to support our present civilization. That too will join the trash heap collectively created by ignorant generations of humanity.

If our collective consciousness is too slow to change to a culture of sustainability then there are drastic consequences to contemplate, which are starkly portrayed in the film *The Age of Stupid* that we have discussed..

Imagine an ancient ecologist on Mars studying a million years of earth history. She would note a parasitic infestation on Planet Earth that was not very intelligent. An intelligent parasite would ensure the good health of the host that supports it. And so the Martian ecologist would factor in an inevitable elimination date for our species in her star-date log and may well view our civilization as a failed genetic experiment.

We may have to accept this Martian musing as a potential reality staring at us from the very near future. Our present values and patterns of consumption are the architects of the present global ecological emergency and we remain ignorant of interconnectedness of the world.

We are, in fact, our environment. It is our collective habits, thoughts and patterns that have created a flimsy, uncertain future for our species.

Every authoritative body on the planet provides dire warnings to humanity about the effects of climate change. We have fixated on external climate change searching for technological fixes.

Yet, climate change is merely the symptom and outcome of a maladaptive human mind-set. It is clear that our current non-sustainable energy and economic systems are not working.

However, policy makers who rush to find alternatives to fossil fuels do so without addressing the root causes of the problem. Pathological consumerism is the major manifestation of industrial civilization and it will bring us all down.

Because of all the warning signals, allow me to be starkly realistic. If the failsafe in consciousness does not kick in, the field is open for James Lovelock's conclusions to take root.

But perhaps after all the Arctic Circle may not be such a bad evolutionary staging point, as digital records, carefully preserved as archaeological relics, could provide clear guidelines for future civilizations to conduct themselves more appropriately with respect to the Earth Mother.

I conclude this essay with Dave Hampton's passionate thoughts about this film (Resurgence May/June 2009: 66). "*The Age of Stupid* is not just a film that could change the course of humanity. I hope it will be the catalyst that gives us a second chance to create a sustainable future. I hope it will promote a mass collective awakening globally so that we are not stupid and that we choose life and reclaim our children's birthright, the right to expect a future."

I have fourteen grandchildren. In the same vein as this film I wrote *Failsafe: Saving the Earth from Ourselves* to provide hope and an action plan so that my grandchildren can enjoy a habitable planet.

Should the adversity of Climate Change overwhelm humanity – then a different question arises. What will we choose as a paradigm of behavior?

Glance at the sun

See the moon

And the stars

Gaze at the beauty

Of the Earth's

Greening

Now Think

- Hildegard of Bingen 1098 - 1179

PART TWO:
FAMILY AND COMMUNITY

> *"I introduced them to the basics of walking meditation, slowing them right down with breath, guiding them to release their distress into the earth. I still smile when I remember this scene: my punk friends and I walking barefoot in the grass of one of Glasgow's finest private parks, breathing slowly and walking mindfully for more than two hours. We sat on a park bench, fresh with morning dew, and they began to talk to me. As I listened to them sharing heartfelt stories of how they came to be where they were, I encountered a level of deep listening I'd never before experienced. I felt an all-encompassing energy embrace me, my young friends, the park, the lights, and night sounds of the city of Glasgow."*

- Dr. Ian Prattis

4: Punk Palace

I encountered a novel form of parenting in the adventure with my teenage son in Glasgow's drug world. It became clear that when all else fails there is still mindfulness and it can work miracles.

In the midst of squalor, alienation and despair, my son and I found humour, goodness and wonderful surprises. It was the practice of mindfulness that allowed me to remain steady, for the most part anyway.

It took hard experience for me to discover how interconnected I was with everything, even with situations I did not readily understand. I realized afterwards that if I can stop discriminating against others, I can know wholeness. This is a life experience, not an intellectual construct.

These insights were brought home in a totally unexpected way. My preparation for "The Heart of the Buddha," a meditation retreat with Thich Nhat Hanh at Plum Village in France, was not at all what I had anticipated. My eighteen-year-old son, Alexander, was studying at the Glasgow School of Art in Scotland, and my transatlantic phone calls that summer revealed that he was deeply in trouble with drugs.

I arranged to spend time with my son in Glasgow prior to the retreat. We had not seen one another for a few years, so a visit was overdue, particularly since he had suffered deeply from his parents' divorce. At Glasgow airport I scarcely recognized him, as he now sported a shaved head with all the required black accoutrements. Yet he greeted me with a warm hug and a big smile.

On arriving at the place where he was living, I knew something was dreadfully amiss. There were no books or art materials in his room. His large rambling apartment was occupied by "The Tribe," a shifting population of punks, drug users, and dealers.

As I sat in Alexander's squalid room wondering about him, he left for a while. There was such an atmosphere of hopelessness that for a moment I felt momentary despair, and did not know what to do.

I found my own deep silence and meditated so that I could be clear and calm. I needed support from all the tools of mindfulness, particularly deep listening, to remain steady and not be drawn into judgment and discrimination. My hope was that mindfulness and meditation would enable my actions to come from the steady consciousness of my heart. The despair slowly receded, and along with it my judgment of his living space.

Alexander returned several hours later, badly beaten up from a drug deal that had gone wrong. He admitted that he was a drug dealer and had lied to me over the phone. His requests for financial support had nothing to do with completing summer art courses. He was deeply in debt to the Glasgow drug underworld.

I listened to him very quietly, stayed calm, washed his re-arranged face, and learned that he could easily have been killed. He had not fought back when beaten by the dealers' protectors. He simply took the beating, perhaps the first smart move he had made for a while, since it certainly contributed to his remaining alive. I needed to be smart about the moves to turn this situation around.

We went for a walk to the nearby Kelvingrove Park, where I introduced him to walking meditation, teaching him to trust Mother Earth to absorb his pain and distress on each out-breath.

As Alexander became calmer, I told him that perhaps the beating was fortuitous, a stark wake-up call about the life he had chosen, and one that coincided with my visit. I offered him two alternatives: a thousand dollars in cash, so he could enter the drug world in a bigger way and likely end up dead within six months, or spending the next few weeks living mindfully with me, so that he might see the difference between what he was doing and what he could be doing. Both alternatives were equal in my mind. I did not discriminate between them. He refused the money, so I will never know how much bluff and shock I had loaded into the first alternative. But that was no longer important, as a magical time of living mindfully unfolded between us.

I had Thich Nhat Hanh's book *The Miracle of Mindfulness* with me. We read most of it together and did all the exercises. I invited Alexander to join me in walking and sitting meditation, to enjoy silence during meal times, and to use his breath with awareness.

I focused a lot of the breathing exercises on martial arts training. I taught him to co-ordinate body movement with in-breaths and out-breaths and how to defend himself. His comments were frequently very funny.

"Hey Dad, this breath stuff is cool – I'm getting a buzz off breathing."

I pointed out the obvious, that the breath "stuff" was both cheaper and safer than taking drugs.

We discovered that we enjoyed one another's company. An important turning point for Alexander had to do with my own substance abuse. He had memories from when he was younger of my drinking too much. During a time of great unhappiness, as my marriage to his mother came to an end, I had used alcohol to cover up the pain and disappointment of a failing marriage.

Everyone in my family had suffered from the marital breakdown and divorce. That I no longer drank astonished him, and I realized later how important it was for him to see this. I also shared with him my own vulnerabilities and struggles. Step by step I had come to choose a way of life distinguished by a commitment to living mindfully and teaching meditation, but it had taken many false starts to get there. He liked the idea of false starts.

Our working meditations included simple things like mindful laundry and cleaning up his living space, which I had dubbed "Punk Palace." I was convinced that the bathrooms and kitchen contained alien life forms and varieties of mould unknown to science. I would walk each morning to a nearby swimming pool for a swim and a shower, and then return with healthy breakfasts for "The Tribe" - fruit, juice, cereal, and honey. One morning I returned to find two members of The Tribe cleaning out a bathroom. They proudly announced that I did not have to go for such early morning swims anymore. This bathroom was for me and nobody else would be using it. I thanked them for their consideration.

Each evening all the residents gathered in one of the five bedrooms to sit, listen to extreme heavy metal music, do drugs, and talk. Alexander had given me a commitment not to take drugs during the time I was there, so he would smoke cigarettes. I listened to these young people as they poured out their lives. I did not judge them and for the short time I was in their midst they became family. No other parent had ever visited them, let alone lived with them. I chose to be with them as my own family. Several of the residents asked me one night if I would teach them walking meditation. They had obviously been talking with Alexander. I said I would be happy to, as long as they remained drug free for two days. They agreed and complied, which was quite an undertaking for them.

Two evenings later, at midnight, my new friends chose one of Glasgow's finest private parks to do their walking mediation. They found a tree just outside the park fence, boosted me up into it, and instructed me to crawl along a branch that overhung the park. For their part, they were much more agile than I and simply bounded over the fifteen foot railings and then caught me as I dropped from the branch in a less than elegant manner. Once we had picked ourselves up and stopped laughing, I introduced them to the basics of walking meditation, slowing them right down with breath, guiding them to release their distress into the earth. I still smile when I remember this scene: my punk friends and I walking barefoot in the grass of one of Glasgow's finest private parks, breathing slowly and walking mindfully for more than two hours. We sat on a park bench, fresh with morning dew, and they began to talk to me. As I listened to them sharing heartfelt stories of how they came to be where they were, I encountered a level of deep listening I'd never before experienced. I felt an all-encompassing energy embrace me, my young friends, the park, the lights, and night sounds of the city of Glasgow.

This experience totally changed my understanding of deep listening, a mindfulness practice I was very familiar with, but not at this level. On later reflection I could see that I had journeyed through several distinct levels of deep listening. The first (and least significant) level of deep listening was intellectual, whereby I analyzed and scrutinized Buddhist literature on the practice, gaining a conceptual grasp of what it meant within the corpus of Buddhist teaching. Although this was the least significant level of understanding, it was a starting place, which enabled a window to open. Rather than an intellectual practice, I began to realize that deep listening was a consequence of mindfulness practice. This was the second level. Deep listening could not exist alone. I experienced a distinct improvement in my capacity when I realized that walking meditation, mindful breathing, mindful meals and other practices were the necessary ground out of which deep listening could arise, as a flower growing from fertile soil. When such a ground was not there, listening would focus on my own agendas and assumptions, and not to what was being said to me. The simple insight that deep listening could not exist alone was important. This belief deepened as I investigated how it worked for me and directly affected my life. There were times when I was not heard and I suffered from them as well as causing equal pain when I was not in a place to deeply listen to the concerns of those speaking to me, especially to my children. I think about my children when they had really important things to say, but I was too busy to listen. I did not stop to give them my full presence. Many years later, now that they are all grown up, I have said to them individually: "I remember the time you said such and such to me and I did not really listen to you. I am very sorry." They were astonished and very deeply touched, and so was I. The rifts between us could heal.

On this evening in one of Glasgow's finest private parks I encountered a much deeper level of deep listening. I said very little and left intellectual understanding and personal suffering behind, entering a whole new territory. The carefully constructed sense of self just dissolved and the "I" of me disappeared. In that moment I was deeply present with my young friends. "I" became like particles of energy, touching and engaging with the particles of energy in everything there - my young friends, the grass, trees, park bench, city lights and sounds, and beyond to a vastness that I cannot find the words to express. In that stillness, the same energy touched the deep seeds of consciousness in my young friends as they trusted me with their confidences and secrets. We stayed there for hours, frequently silent and walked back to Punk Palace just before dawn. From the smiles and embraces we exchanged

I knew that something had changed in all of us. I had discovered a level of deep listening I had never thought possible; my young friends and son had nurtured long forgotten seeds of hope within themselves. My gratitude for the gifts received was enormous. My journey and practice of deep listening had travelled from an intellectual and experiential level to an instrument of transformation. In addition, interconnectedness for me was no longer just a good idea. It was a direct experience of reality.

My new friends now showed great consideration for me. Drugs were still being used, but less so. When I was there, they'd turn their heavy metal music down, as they knew I'd not yet learned to appreciate it. No drug deals went down while I was there and the kitchen even got a cursory cleaning. At the evening sit following the adventure in the park, they all said they were very aware of my presence in Punk Palace and quite liked the name I had given their abode. I thanked them for their consideration and quietly said I was very much aware of them and their consideration of me. I was also aware of every acid hit, of every cocaine use, and of every moment of their despair, anger, and self-destruction. I felt the energy of it all in my body and it hurt like hell. A silence ensued that dragged on. One of the girls and one of the boys started to cry. Yet it was a good silence, for it had healing and heart in it. I broke the silence by very gently thanking them for their kindness toward me and told them that I was there for all of them. Then I left them to talk amongst themselves. They had listened to my stories of wilderness adventures in Canada, of my pet wolf and how it felt to swim with dolphins. They instinctively knew for themselves how everything was interconnected. They were simply lost.

After that evening, I did many walking meditation exercises with each one of them in the nearby park. I spent time listening deeply to them and learned a great deal about the angst of alienation amongst young people. I was there to provide counsel when asked and always steered conversations around to the topic of taking responsibility in their lives. They felt abandoned and marginalized, yet were so creative and intelligent. They talked openly, for they sensed that I did not judge them. I fed them with healthy food and counseled them in simple terms that related to their situation. The basic frustration for my punk family was the powerlessness they felt over their lives, a deep hopelessness they escaped from through drugs, pimping, prostitution, and drug dealing. Each time this would arise in our many talks and walks, mostly one on one, I referred to a fundamental understanding: power over one's life comes with taking responsibility. We talked about their creative dreams, the dynamic energy they had neglected, and practical alternatives to the way of life they had fallen into. I talked about my own struggles in life, including dealing with the childhood sexual abuse that was buried in my unconscious mind for many years. That was a point of connection, for most of them had a history of abuse and neglect of one kind or another.

My working sessions with Alexander continued and involved practical measures that we prepared for by practicing meditation. We met with his college tutors, who had not seen him for six months. He told them about his drug activities and was surprised to discover that his tutors were fully supportive in providing guidance and tutorials enabling him to redo his first-year courses. He voluntarily entered into drug and alcohol counseling. Alexander had been very creative with his bankcard and overdraft to finance his drug activities, so his bank manager was quite amused by the course of our joint meeting. I cleared his deficit then asked Alexander to give me his bankcard. With a pair of scissors supplied by the bank manager I cut his card up and instructed the manager to withdraw all overdraft privileges until she was sure he'd be responsible.

Alexander was astonished, exclaiming, "I don't believe you did that Dad!" Yet he told me later that he admired the firmness and clarity. I also enrolled him in a martial arts academy, as he needed a safe place to leave his frustrations and anger. It was run by a rugged international kick boxing champion who also had a wonderful heart. Many years earlier, I had been one of his Canadian opponents and had fought him several times. I was impressed by the quality of his instruction, by the way he treated his students as an extended family, and also by the fact that his training sessions began and ended with meditation exercises.

The final step was to talk to the drug dealers. I met with some of them in Alexander's room at "Punk Palace." I had expected to meet Mafia-type figures. After all I had seen the Godfather movies. But Alexander introduced me instead to young people who were hardened to a degree I had never before encountered. I cleared Alexander's debts with them, and the message from me was quiet but firm. Alexander would not be doing any deals with them anymore. You could cut the tension in the air with a knife. I made myself breathe slowly in and out and extend kindness and compassion to them from my heart. That was all I had. They had guns and knives. I only had breathing in, breathing out, and deep listening. After a time, they too relaxed and asked about my martial arts background, which Alexander had no doubt exaggerated. I'm not in Arnold Schwarzenegger's "Terminator" class, but they had the misguided impression I was. A false perception, I should add, that I did nothing to correct! This was fine at the time, as it was the only common ground, apart from Alexander, that these hardened young people had with me. I wove a web of stories and when asked, showed them several quite deadly moves. Eventually I spoke to them about the many martial arts experts, such as Bruce Lee, who had a base in healing and meditative practices. Not a seed that may get watered in their world, but at the time it was the best I could find. The more I talked quietly and directly to them, the more the violence left the room. In the end I was silent and Alexander did the talking. When they left, I knew they would leave Alexander alone, as there was that unspoken "honor" that sometimes arises in these situations. Nevertheless, their energy greatly disturbed me.

It would be ideal to say that all of this did not really get to me, but that would be untrue. After one all-night party at Punk Palace with acid hits flying and heavy metal ruling the airwaves, I got really angry. Several sleepless nights did not help and although I knew Alexander was not using drugs, I was angry at his pattern of wasted opportunities and irresponsibility. I looked deeply into this anger and saw shades of the same patterns in my own past. I tried walking meditation to calm myself, so I could respond rather than react but it was not working for me. I was still angry so I dressed and packed my bags at 5:00 a.m., found Alexander and asked him to walk me to the bus stop. I was leaving. His shock and panic were palpable, and the fear that I was walking out of his life showed on his face. We walked in silence to the bus stop. Alexander insisted on carrying both my bags, which were much too heavy for him. But I let him do it anyway. Then I stopped, asked him to put the bags down and hugged him. I told him I loved him. We were both crying as I explained my anger. And do you know what he said? He told me the party was for me, but they all thought I was sleeping. I had to laugh at that one, whether it was true or not. I made it clear to Alexander that I was there for him but that I had limits. I invited him to join me at the airport hotel for the next few days to continue our mindfulness training together. Relief flooded his face and he apologized for not considering my feelings. I apologized for getting so angry with him. I had suddenly and clearly seen what his life had been this past nine months and was upset by the wasted opportunities that irresponsibility brings. We both cried again.

That evening after his kick boxing class, Alexander joined me at the airport hotel and our mindfulness training continued with an emphasis on life skills: how to budget one's money, handle peer pressure, complete college assignments, do research and so on. We meditated a lot together and continued the breath work with further martial arts training. Once again we drew closer. Then I left for the retreat with Thich Nhat Hanh. Alexander saw me off at the airport and the real test, both for him and for me, began. He had to choose how he wanted to walk through life and I had to allow him the freedom to choose.

At Plum Village I shared this story with the Lower Hamlet sangha. My friend Sister Vendana, an elderly Christian nun from India, asked the whole sangha to hold Alexander steady in their hearts and everyone did so in a deep and beautiful silence. I ask the reader to stop now, and take the time to hold in their heart not just my son Alexander, but all young people who are lost. Would you please do this? Would you hold all lost children close to your heart and extend the energy of loving kindness to all of them wherever they may be?

Meditation for Lost Young People

(Bell)

Sit quietly and breathe in to the heart center in the middle of your chest, and up to the crown. Pause for a moment. On the out-breath take the light of the universe, and of your heart, down through your body from your crown to your toes. Continue to breathe in and out like this for several minutes until you settle into a steady calm. Once you feel calm:

On the in-breath say to yourself: I am aware of the suffering of lost children;

On the out-breath say to yourself: I am calm enough to deal with this suffering.

Continue to breathe in and out like this for ten breaths until you feel stable and solid.

> IN-BREATH: AWARE
>
> OUT-BREATH: CALM

(Bell)

Once you feel stable and solid:

On the in-breath say to yourself: There is love and healing energy within me;

On the out-breath say to yourself: I extend this energy to all lost children.

Continue to breathe in and out like this for ten breaths until you feel an expansiveness and sense of interconnectedness within your being.

> IN-BREATH: ENERGY OF HEALING
>
> OUT-BREATH: EXTEND TO LOST CHILDREN.

(Bell)

Breathe in and out with this awareness and direction of healing energy for approximately ten minutes.

(Bell)

Slowly open your eyes and consider if there are ways and means for you to help lost children in your home community.

During the Plum Village retreat my thoughts frequently turned to Alexander, yet I was calm and surprisingly confident, as I do believe that mindfulness brings about miracles.

Two weeks later I received a wonderful letter from Alexander.

Here are some excerpts:

> *Dear Dad,*
>
> *I hope this letter finds you well and at peace. How is Plum Village, Thich Nhat Hanh and the early morning schedules? I know I'd be hard pushed to get up early for more than two days in a row......*
>
> *If Plum Village is anything like I imagine, it will look and feel amazingly beautiful and tranquil. One day I hope to have the opportunity to travel there and experience what it has to offer and hopefully bring whatever I have to it. However, right now and for the next few years I'll be busy with my studies ... I hope for you all the best and would like you to pass on a message to Thay for me. Tell him that what I have learned and discovered from your guidance, and the reading of Thay's letter to a brother monk written many years ago (The Miracle of Mindfulness); is that I have made the shift from a poisoned, lost and confused kid into manhood. Although I am still testing the waters in places, I have this overwhelming desire to better myself. I hold on to the words and example that are close at hand to guide me through the rough spots. Thank you both, Dad and Thich Nhat Hahn. I am very, very grateful.*
>
> *Lots of love,*
>
> *Alex*

I was very happy. I shared the news with my friend Sister Ani Lödro, a Tibetan Buddhist nun. She had earlier expressed a wish to present a tree to the Lower Hamlet, to commemorate the beautiful and aware death of her friend Kay. We agreed to join forces in this enterprise, so I could add a commemoration for my son's new life.

We purchased an apricot sapling from the nearby garden center at Bergerac, and of course it grew bigger than our immediate and particular concerns.

It became a sangha building tree that belonged to everybody. It was planted in Lower Hamlet and Thich Nhat Hanh graced the occasion with his presence, placing the first earth around the tree.

The Lower Hamlet community, drawn from all over the world, gathered together for this tree planting ceremony.

Each person placed soil around the apricot sapling and added their own heartfelt prayers as we committed this symbol of community to the earth.

There were prayers for the starving children in the world, love for the planet, and songs for world peace, guidance for our leaders to be wise and compassionate.

A Soto Zen priest from Japan played his flute and then sang an earth shaking chant.

My friend Mike Bell from England was holding the tree as everyone paid homage in their own unique way. I saw that there were tears in his eyes, as in mine, as we witnessed the Plum Village sangha making a commitment to be parents to the world, each in their own beautiful way.

A sangha that I thought of was "The Tribe" from "Punk Palace" and my prayer was for all of them to create the opportunity to walk mindfully through life.

I did not see any of the tribe until several years later, as I was travelling to India after the Plum Village retreat. I was amazed at the steps they had taken, the responsibility they had exercised in different ways.

In the midst of Punk Palace and the Glasgow drug world I had learned firsthand that when all else fails there is still mindfulness and it works.

Furthermore, that interconnectedness is no longer just a good idea. It is a concrete experience of reality. If what we seek cannot be found in Punk Palace, it is doubtful we will find it at all.

Alexander's progress was not unlike that of many young people, two steps forward, one step backward.

Yet, the alternatives were very clear to him as he began to place his life within a different orbit.

Several years later it took him and his girlfriend on a journey to Vietnam and Cambodia, to Buddhist sites and temples. They now have two beautiful daughters and the responsibilities of parenthood sit well with them.

The seeds from our adventure in Glasgow continue to be watered.

One of the most memorable occasions that I can bring to mind from this unexpected adventure in parenting was when Alexander and I sat on the doorstep of "Punk Palace" watching the full moon rise above the city of Glasgow.

We took turns composing lines of a poem to the moon and I cannot discern where he began and where I ended, which is perhaps just as it should be.

The collaborative poem composed by my son and I, *Punk Pace in the Moonlight*, can be found on the next page:

Punk Palace in the Moonlight

The moon a delicate mistress,

veiled by fleeting clouds and mysteries,

reveals that penetrating light of grace.

It makes the stars and galaxies dance.

The moon does this

with all that is in me.

This gateway to boundless space

is the door to my troubles and joy,

For I am in the moon and stars

and they are in me.

We dance together

- Now Bright; Now Turbulent

Now Lost; Now Found -

Beyond sense or reason.

And the night sky casts movement and hues

To something I touch

with that in me.

The full harvest moon

rises from banks of pastel grey,

pacing existence

through the rhythms of our universe.

5: More Dead Children

The specter of children shooting children in high schools shocked North America, yet very little institutional change has been effected, once the platitudes of politicians receded. This chapter examines the consumption of violence by our children through the media, video games, and internet. It can lead to the deadly carnage of high school shoot-outs and murder, particularly when mental illness is considered.

Young people, their parents, and society at large are unaware of the necessity of guarding their sensory doorways and mental health. I illuminate the very dangerous environment we have created, and offer practical measures of mindful engagement as a way out. Young people need simple tools to deal with their hate, anger and distress so they do not resort to guns.

The shock waves and horror of the 1999 high school shootings at Columbine High School in Littleton, Colorado and the 2012 massacre at an elementary school in Newton, Connecticut swept across North America and touched every community.

Since then, massacres in schools and universities have become more common. As these shock waves receded, the greatest danger is that the public may distance itself from taking responsibility for the toxic environment it has created. High school students across North America, however, have not forgotten. On the anniversary of the shootings at Columbine High School, many students across America refuse to go to school for fear of a repeat shooting spree. Their fear is that "IT" could happen in their school.

Personally, the specter of children shooting children in high schools shocked me deeply. I was offended by the carnage and angry at society for creating the causes and conditions for children to end up murdering other children.

I also had meditation students who had settled in Colorado, and they phoned me in a panic. I knew I could be of little help to them, for I was not in the appropriate space to give counsel to anyone.

I first had to find the bedrock of understanding and compassion within myself before I could communicate anything worthwhile to others.

To come from anger and shock was not something I was prepared, or trained, to do. I requested that my friends focus on Walking Meditation, to calm themselves and others around them. I would get back to them once I had taken care of my own anger and distress.

After three days of silence and meditation I wrote this essay-chapter. I looked into the causes of the shootings, and saw that with the passage of time people would become removed from any sense of personal responsibility. There is the ready availability of guns and drugs that readily collude with the problem of untreated mental illness.

When individuals are raised without influence from parents, teachers and community leaders, the consumption of violence can become a large influence in their lives. It is readily available through the media, video games and the internet, influencing those involved in shoot-outs.

Many of our children have become exiles, their voices unheard, and we have largely forgotten how to listen to them. Some find a place in cyberspace where violence, hatred and killing are readily available without any sense of consequence or responsibility.

In the absence of clear ethical guidelines from parents and society, young people are creating their own identity from the very worst that media has to offer. This identity can be built through a drive to achieve instant fame through acts of notoriety.

Children who have built positive core values through the influence of parents, teachers and community leaders have an internal strength to resist this seduction.

But children who have fallen through the cracks are without support and guidance. They live out their sense of exile through the cruel fantasies available to them and become desensitized to the consequences of violent acts.

I write about dead children and not just the twenty killed in an elementary school in Connecticut and elsewhere in North America, but for children killed as collateral damage in world-wide violence.

We are all grieving parents to the world. The question is: What Now?

In the face of grief we must feel it deeply, be hurt by it, taking time to feel the pain of the tragedy. Then come through, determined to make a difference. This requires calling in the support of wise friends, counselors and community so we can begin to see clearly and find ourselves.

Stillness is needed, not social media distraction and drama, for we need a new direction from leadership. To reassess the 21st century, we must look deeply at the factors involved in the shootings.

In the United States there is a complex, intertwined tapestry with the easy availability of guns and drugs. This is compounded by societal tolerance of violent media, plus the very serious common denominator shared by the killers stretching back to the Columbine massacre.

The self-delusion and mental health issues in predominantly pre-adult white males. They are caught in an identity trap that they escape from through violence and murder. Through killing, they gain five minutes of fame that enables them to be remembered. They occupy a toxic landscape of "not love," "not connected." And this is what requires the attention of our health system and mindfulness. How do we begin?

It is time for the Bodhisattva to enter the 21st century as a paradigm and archetype for individual and collective action. This enables us to transform ourselves and our civilization.

We nurture this paradigm by cultivating two aspects that lie dormant within us. The first aspect is interconnectedness, knowing that we connect with everything, the earth, oceans, forests and mountains, all species and most of all – with all people.

Fostering interconnectedness creates harmony and unity and destroys the ego. The second aspect is non-discrimination, which carries the energy of compassion, and this combination threatens selfishness. Taken together, these buried aspects manifest from within us, opening pathways and bridges to build a better world.

How do we do this? We cultivate the energies of transformation: mindfulness, concentration and insight. At every opportunity we bring interconnectedness and non-discrimination to the forefront of our daily lives. In this way we shape the future of the 21st century as we begin to live differently.

We should not be intimidated by present crises. We are certainly shocked and hurt by such circumstances but are much stronger than we think. Enter the Bodhisattva is the guiding paradigm for our lives.

I allude to Bruce Lee's classic, *Enter the Dragon*, which brings the fierceness of the warrior to the fore and the determination of a saint to overcome tragedy and set a new course. It takes practice, skillfulness and creative vision. Are we equal to the task?

Most young people, their parents and society as a whole, are unaware of the need to guard their sensory doorways, or eliminate their engagement with violence. One can easily see that violence in the external environment must be controlled.

As an alternative, steps must be taken in schools and communities to deal with frustration, mental health issues and hatred without resort to guns. Our senses can be bombarded by so much damaging material. The violence that pours in feeds the consciousness that drives us. If we load our mind with toxins and violence we should not be surprised by what occupies the driving seat.

My body and mind are not individual entities that I can do anything I like with, such as filling them with drugs, alcohol, hateful attitudes, harmful identities, unhealthy foods, cravings or violence. My body and mind exist to provide for future generations therefore I must be aware of what I put into them. We also must exercise care and responsibility over what we allow into the minds and bodies of our children. This care and responsibility prevents young people turning their consumption of violence into violent acts on themselves in the form of suicide.

We must say "no" to our children consuming violence through movies, video games and hate concerts. At the same time we refuse to engage in violent and toxic interactions with them. We must take steps to fill the ethical void and give our children the benefits of our full presence and learn to listen deeply to them.

But when was the last time anybody really listened to you? And when was the last time that you really listened to your children? Our listening is usually filled with judgements, and young people are deeply hurt by this. To listen requires that we find a way to leave our judgements behind, to be present.

We may understand our children if we listen compassionately. When we are fully present, our energy can transform them and heal their deep hurts, erasing neglect. We learn about full presence through meditation.

The teenagers who murdered their classmates at Columbine High School had no-one to listen to or be present with them. Nobody helped them or took care of the violence that flooded their consciousness.

The Rev. Dale Lang, who lost his son Jason in the high school shooting in Taber, Alberta, provided a wonderful example of leadership and forgiveness for his community, in the midst of his own personal grief.

He asked that his son's death not be in vain, and that the community forgive the boy who killed him, that they practice compassion. From the families of children killed at Columbine and in Newton, there is the same plea. Let their deaths not be in vain. We can respond by recognizing that we are either part of the problem or part of the solution and must examine how we support and condone the culture of systemic violence.

If time passes and nothing changes, if we sit on the fence and say "this is not my responsibility" then we are part of the problem. There are many parents, teachers and community leaders who are endeavoring to make a difference to the existing state of affairs in their homes, schools and communities, but their efforts may be too slow for our children.

The high school murders are not a teenage problem. They are a societal problem of systemic violence penetrating to the consciousness of young people through their sensory doorways. Thus a societal solution is necessary. One that deals with anger, frustration, mental health and hatred. It must also provide an alternative paradigm that impacts the internal environment of violence and transforms it.

One reason there are disturbed young people is a lack of positive role models. Neither group had adjusted in order to protect themselves from toxins and violence in the media. If we do not guard our sensory doorways, there will be negative effects.

After the platitudes of politicians and the media were delivered following the high school and university murders, not much has changed in terms of institutional structures or constraints on the production of violence in the media, video games and movies. Providing young people with tools is essential. I suggest mindful engagement, the subtitle of this collection of essays.

There are parents and teachers everywhere who are desperate for a change of direction, recognizing the enormous crisis. The Chinese letter for crisis has two characters, the first is danger, the second is opportunity. We need to recognize the danger of violence combined with mental illness and seize the opportunity of mindful engagement to deal with it.

In the space created by meditation, the toxic and violent consumption of everyday life has no doorways to pass through. It is not a total solution but it is a start.

To young people I recommend simple tools of handling anger, hatred, distress and mental illness. There are techniques of meditation that allow the person doing them to recognize their negative emotions and deal with it.

I tend to think that the most useful technique is walking meditation, simply because it is hard to sit and meditate when you feel angry or violent (further discussed in Chapter 6).

We can literally walk ourselves out of crisis by taking care of the distress and releasing the energy of it into the ground. Recognize the danger of anger and hatred and seize the opportunity of walking meditation to deal with it.

I invite meditation teachers to take their skills into schools and community centers.

Providing these methods can make an impact on the anger and hatred that affects our children. I invite young people to bring such teachers into their midst and see what they can teach.

We are either part of the problem or part of the solution. As citizens we all have the capacity and the responsibility to change things for the better: at home, at work and in our public life. We have to take a stand not to condone violence.

It is our actions, from a space of clarity that provides solutions.

Our indifference to the dangerous environment we have created means that we perpetuate the problem. I ask everyone to choose wisely, and immediately.

This message has been sent far and wide, thanks to internet technology, and to good people everywhere who passed it on through their own networks. It was used in many communities, particularly Colorado.

In my own city, Ottawa, I gave workshops and retreats for students about violence in schools and mindful engagement and continue to do so.

6: Cyberbullying in Schools and Teenage Suicide

Some triggers for teenage suicide were brought to my attention through a drastic and dangerous situation with one of my young friends. He had slipped into a deep depression, primarily caused by being bullied at school and was seriously contemplating suicide.

His father had phoned me in alarm and I suggested that his son come and stay with my wife and I for a while. I also had a long conversation over the phone with the young man without mentioning the word "Suicide." I talked to him about our kayaking adventures with his dad and other things that I knew would bring some joy and happiness to his mind. These were the first steps to transform the feelings and emotions that led him to consider suicide.

Over the phone, I also taught him a simple meditation about being a tall tree. He was open to Buddhist "stuff" through my and his dad's practice and was somewhat curious about both of us. The analogy I used for the meditation was that of a storm of violent winds coming up and shaking the tree tops and breaking branches, while the bottom of the tree trunk stays solid.

"My dear young friend, think of yourself as that tree and the violent storm as the anger, hatred and despair that overwhelms you at times. If you stay in the tree top with your mind and your reactions, then surely something will break. This is very dangerous for you can lose it, say and do things that can harm others and yourself. Notice the lower trunk of the tree that remains steady in the midst of the violent storm. Now place your two hands flat on your belly, below your navel. As you breathe in, say to yourself - "I am aware of breathing in deeply to my belly."

I added: "As you breathe out, say to yourself , "I am aware of breathing out slowly from my belly." Do this for ten to twenty breaths and feel the calm settle in, and notice the storm of emotion, anger or frustration is not so strong. You're now in a position to act without violence towards yourself. Try this the next time you feel overwhelmed."

"Yes I will. Thanks." He then accepted my invitation to come and stay with us. I also learned that the youth was being most injured by cyberbullying from anonymous sources.

I had no knowledge or insights about cyberbullying. Before he arrived in Ottawa, I consulted with school councillors across the country. I learned that cyberbullying was now an everyday reality for teenagers in schools. I was shocked by the ramifications of cyberbullying and by the fact that a whole generation of school children had grown up with it, becoming an everyday mosaic in their lives.

It has become a forum for negativity without restraint, simply because people can remain anonymous. The impact on victims was very severe leading to breakdowns, depression and sometimes suicide. Parents and teachers were often completely unaware of this. With this new, alarming knowledge I knew I had to share some simple practices, at least to get my young friend started.

My wife and I picked him up from the airport in Ottawa and made him completely at home. At first there was no mention of his depression and suicidal thoughts. My wife fed him mounds of food. It seemed he emptied the fridge at least twice a day. He'd sleep as long as he needed to and rest. Sometimes he'd join me in the meditation hall in the basement of our bungalow. He was curious about my practice, so I taught him how to make good friends with his breath, concentrating on the whole length of the in-breath and the whole length of the out-breath.

I said that if he would do that ten times without distraction he would feel calm. He also joined in when I did walking meditation. Here the breath was co-ordinated with each footstep and an amusing mantra to follow each breath.

Say IN - OUT, with left foot and right foot.

Say NOW – WOW, with left foot and right foot.

He smiled at that. Furthermore, when I added the final concentration of being aware of how our feet touch the floor: heel-ball of foot-toe, he could in fact align himself with Earth Energies. I told him that this part of walking meditation was very important, as it was the catalyst for the strong earth energy already inside his mind to come to the surface. And this energy was stronger than his troubled feelings and emotions. He looked at me carefully as I provided a demonstration. When he practiced it, he found it to be okay. He related, much later, that walking meditation was the best for him, as he felt a sense of steadiness and of being refreshed.

"Walking meditation is a good start," I said. I explained that when we concentrate on our breath and focus on slow walking, we have a brilliant piece of engineering to quiet the mind and body and be clear. When we add the third concentration of being aware of how our feet touch the earth, we have a meditative practice for our troubled times. We slow down even further and with our body, not our intellect or ego, make a contract with Mother Earth to leave a smaller footprint. We can then examine our consumption patterns and energy use with clarity.

Over the two weeks he stayed with us his visits to the meditation hall were intermittent but by the second week he came down every morning in his pyjamas to keep me company in the meditation hall.

Once he got dressed each morning and after a late breakfast, I would take him to the various science and technical museums in the city, as that was his passion along with First Nations culture. In Ottawa I knew several curators, one at the Aviation Museum and one at the Museum of History, which had the Grand Hall of North West Coast Cultures.

My friends in the museums kindly gave him individual tours. I could see his sense of self-esteem rising with the tours and kindness. He was over the moon about receiving such special attention, gladdening his mind.

This prompted me to think deeply about what practices would be useful to ground the troubled minds of teens so they could resist cyberbullying and prevent being pulled into self-hurt. I had to be selective and intelligent about mindfulness practice. Strategic too, so it would be readily grasped by a young teen. It was clear to me that cyberbullying was a malicious enhancement of unworthiness and hate. Many teens played both sides of this virtual reality, victim and bully, so rampant and vicious was this spectre of hate.

I started to talk to my young friend about foundation practices I used every day and how they might help to calm his mind when he was troubled. He connected with the Two Arrows Teaching from the Buddha. This teaching is about a man walking along a path when suddenly he is hit by an arrow fired by a hidden and unknown attacker. The pain was terrible. Then a second arrow was fired into the same spot and the pain and suffering became unbearable. I asked him if he knew who fired the second arrow.

He slowly nodded his head and said, "That would be me. All my fears and insecurities would come up to inflame the hurt of the first arrow."

I was impressed. I told him he was exactly correct, that our fears, anxieties, exaggerations and dramas inflame the first wound, causing a small ember to explode into a raging forest fire. The point of the teaching was to come to a stop, to calm the mind and body. Then find a way to not fire the second arrow. Buddhism was not such a drag for him after all.

We played board games together and kept on gladdening his mind. He meditated with me quite often and each time we would do the tree meditation together. When the time felt right I asked him if he would like to talk to me about what was going on. He told me about three boys who bullied him at school. He also felt that they were behind the cyberbullying, although he had no proof. Furthermore, that his parents rarely ever listened to him. I listened quietly until he finished talking. Then I picked up the telephone and found the number of his school and talked to his vice-principal for a while. She was very open and supportive and had already taken steps to separate the three bullies, keeping two in detention during every recess. I also telephoned his parents and reminded them about deep listening, which they promised to put into practice with their troubled son. This boy had listened to the phone calls and was amazed at the support for him that was being galvanized right before his eyes.

I also brought to his attention that his father and mother were deeply worried and doing their best for him. That if he decided to "off" himself, his father, mother and little sisters would be devastated. He genuinely did not want any of that to happen. We talked about emotions and feelings overtaking us. He understood that he was letting one or two strong emotions get him down, when he had so many others to choose from. I convinced him that his feelings and emotions were not fixed. They are self-created in his mind by triggers. That we sort of make it all up as we go along and often increase the impact of triggers. The trick, I told him, is to notice when we are getting stuck on one or two heavy emotions. Then we must ask ourselves, "Do I want to go there, knowing what it will lead to?" I repeatedly emphasized that with this kind of awareness he can begin to stop harming himself. He really got this. His understanding was that triggers such as cyberbullying were a spark. He could either stamp it out or create a raging forest fire. He had turned the Two Arrows teaching into a personal tool and clearly understood the difference between responding and reacting.

I introduced him to the mindful use of breath. The focus was on calming his feelings, emotions and mind. Then we spoke at length about how we all love our dramas and allow ego-distortion to run rampant with our feelings and emotions creating all kinds of out-of-control reactions.

However, if we can catch our dramas fuelled by painful mental formations, we can do an end run around our suffering by not firing the second arrow into our pain. We can go deeper and learn how to respond rather than react. We see how our feelings can actually condition the mind. Feelings are totally normal. It is simply a matter of having the stability of mind not to be overwhelmed by them. So why allow one or two feelings or emotions to take us down?

I asked him to write down the main feelings and emotions that drove him to think about suicide. There were three. Then I asked him to write down all the other feelings inside him. He took his time and wrote down thirty. Then I showed him the two figures, three versus thirty.

He nodded his head and remarked "I get it. It's an absurd decision." He understood that the particular emotion that was overwhelming him and making him dysfunctional was just one emotion in his vast ocean of consciousness.

His insight began to undermine the predisposition to be totally crushed by one or two negative emotions. It is important for young people to understand this, as they can quickly go into despair, even suicide, when overwhelmed by emotions of fear and unworthiness.

Deep in our consciousness there exists many positive and wholesome seeds of potential just waiting for an opportunity to manifest in our mind. We gladden the mind by taking conscious steps with our thoughts and intentions, nurturing the seeds of love, compassion, joy, equanimity and other concentrations.

Furthermore, we take positive action by organizing our everyday living so that external circumstances further the nourishment of the wholesome seeds latent in our deep consciousness. I stressed that we have to become very attentive about not dwelling on unwholesome seeds like hate, cruelty, despair, anger, jealousy and greed.

I emphasized to him that nothing survives in our mind without our allowing the flow of nutriments and energy to feed it. He learned to investigate the nutriments that fed harmful notions in his mind. Once we become aware of the causes that feed our negative thoughts we can reduce their potency. First of all recognize the triggers that kept the affliction in our mind alive. Then choose to cease feeding the harmful mental formations by cutting off the nutriments that fuels them with energy. We stop feeding our demons.

The exercises played a big part in in this youth's rehabilitation. The focus placed on gladdening his mind was vital for him to eventually see that he could change his internal monologue. We had listened carefully to him in order to encourage him to eliminate negative thoughts, the 'food' for his problems so that he could stop feeding the nutriments that inflamed his damaging mental formations.

After emptying our fridge one day and finding it bare he started to laugh and said, "Nothing survives without food!" He got it and I was very proud of him and told him so. He had learned about very valuable tools and found some balance and steadiness.

These practices to calm troubled minds are derived from the teachings of the Buddha. They are reliable and relevant to the 21st century realities of cyberbullying. In modern day Canada it is easier to refer to them as mindfulness practices, which can be used for dealing with difficulties, regardless of your personal or professional situation. This stance complements the efforts of all levels of government, from City Hall to the Federal Government to deal with cyberbullying in schools. There are many school boards with apps for anti-cyberbullying, dedicated educators and concerned parents offering their skills to deal with the specter of cyberbullying. This is another very valuable tool in this fight.

On reflection, I saw some key factors that are useful to highlight. They are not offered as a solution to all people dealing with suicidal thoughts from teenagers in schools. The causes of those tendencies are complex and each situation has to be dealt with uniquely.

Nor do I think our practices with this youth can be readily replicated in other circumstances. This young man had a prior exposure to Buddhist practices, as his dad had been one of my meditation students, which made him open to methods of mindful breathing and walking.

The 'treatment plan' my wife and I developed was not in any way formalized. It evolved on a daily basis. Having this youth leave a troubled environment was a great start and consulting with his parents about his home situation was crucial. Surrounding him with love, attention and deep listening was vital.

Teaching him how to remain calm, control his feelings and take back his power through the exercises was an effective strategy. It worked well, as he has since grown into a mature, thoughtful and caring young man. He was prepared to notice the behaviors that devastated him and take steps to try something different. I pray that other teens suffering from cyberbullying and negative situations will be so open.

Cyberbullying is a new phenomenon for our times, scarcely thirty years old. It affects all strata of society, through online forums, list-serves, social media and other internet vehicles providing online anonymity. Cyberbullying can be related to distraction technologies, which create an ecosystem of interruption technologies that many people become addicted to. They crave a global interconnectedness directed by this virtual world, yet rarely know how to use it responsibly.

In less than a generation the world has been fundamentally changed and we have yet to catch up with its consequences. There are not sufficient 'failsafes' and regulations for curbing cyberbullying. Parents and councillors are scrambling to deal with it and parenting skills have to adapt radically to protect our young children.

The bottom line, however, is that distracted people do not realize they are in danger. Neuroscientist Nicholas Carr in his book *The Shallows* (2011) documents a vast amount of scientific evidence that excessive use of the internet impairs human mental capacities. Margaret Wheatley (2013) writes: *"We have made this world into an unpredictable monster because we've refused to work with it intelligently. And the ultimate sacrifice is the future."* There are many other reputable sources bringing attention to this overwhelming issue. We need many solutions, especially as young people see many forms of systemic negativity and cyberbullying in their daily lives.

"Those who are waging war would do better if they knew better, but they don't know better. We, however, do know better and must take the steps to communicate our understandings to political decision makers. It is our developed consciousness, which allows us to know better. It is the meditative work we do on ourselves every day of our lives to come to terms with the inner struggle, turmoil and trauma. This is the inner war, which we must learn to identify as our own and find ways to transform our often raging thoughts."

- Dr. Ian Prattis

7: **Community Activism in Ottawa**

My full time job in the summer is to organize a big event in Ottawa City Hall – the annual Friends for Peace Day. This has been my job for over a decade.

It all started on a bitterly cold winter evening years ago, as the Iraq war loomed. I received notice that a Peace Song Circle was happening on Parliament Hill to protest the bombing of Baghdad. So I went, accompanied by my wife Carolyn, a friend and our dog. No-one else turned up, as it was so cold. I remarked to Carolyn, "This is a good idea but it needs attention to detail and organization."

She replied, "Let's do it." And so we did, creating the nucleus for Friends for Peace Canada. It quickly grew to a loose coalition of over fifty organizations in the city and we asked them to begin the peace process first of all within themselves, then to the community and out to the world.

Our mandate evolved from peace advocacy to projects on the ground. We gave annual Peace Grants to local and international organizations making a real difference, as well as working in concert with other coalitions in the city for environmental and social justice issues. We organized five thousand participants at the Peace Song Circle on Parliament Hill in Ottawa, held on a miserably wet, cold spring day in 2003. A sea of multi-coloured umbrellas on a rain swept morning welcomed all those gathered. As other peace protests joined us and sang "All Within Me Peaceful," the crowd covered the grounds of Canada's seat of government, all meditating at the end in total silence as the rain poured down on our heads.

The pouring rain was strangely welcome, for it symbolized the tears of Iraqi children, my tears, your tears, transformed into hope through singing for peace with one another and experiencing deep peace. There was a transformation of anger, anguish and violence into a determined clarity to be peace and to oppose war. From there we know the wise actions to take.

Those who are waging war would do better if they knew better, but they don't know better. We, however, do know better and must take the steps to communicate our understandings to political decision makers. It is our developed consciousness, which allows us to know better. It is the meditative work we do on ourselves every day of our lives to come to terms with the inner struggle, turmoil and trauma. This is the inner war, which we must learn to identify as our own and find ways to transform our often raging thoughts.

The current projects in the city supported by Friends for Peace include: the Multi-Faith Housing Initiative, the Dave Smith Youth Treatment Centre, Child Haven International, and Peace Camp Ottawa, which brings reconciliation to Palestinian and Israeli teens.

In addition we funded the Physicians for Global Survival initiative to expand the mandate of the Canadian War Museum to include the creation of a culture of peace. There were other projects in Africa, India and Nepal. One planetary care project was the campaign to make the Dumoine River watershed in Quebec a protected conservation park. Peace Grants were also awarded to rebuild the Galai School in Liberia and the Healing Art Project of Minwaashin Lodge – an aboriginal women's centre in Ottawa. Orkidstra received the 2011 grant to expand their children's orchestra. 2012 Peace grants were presented to Dandelion Dance Company and USC Canada.

Ottawa Friends of Tibet received a Peace Grant for their Tibetan Re-Settlement Project in 2015.

Each year since 2003's relentless rain on Parliament Hill, annual Friends for Peace Days have been memorable. We were rained and snowed on for several years on Parliament Hill, thunder and lightning at Alumni Park of Carleton University, before we moved inside to Jean Pigott Place in Ottawa City Hall. We organized differently there, with peace activist and environment booths along the periphery of the hall, a food court at the back, a long set of tables with items for the silent auction and the stage at the north end. The response to this community activism has gone beyond any of our expectations.

Here are some highlights:

*Peace Award recipients Bonnie Cappuccino (2004) and Bruce Cockburn (2006) in their Vision speeches ripped into government foreign policy for neglecting global warming. Dr. Peter Stockdale (2010) held City Council to account for inter-ethnic violence and neglect in Ottawa.

* In 2010 Clive Doucet, candidate for Mayor, not only danced a great number with Big Soul Project, he gave a stirring speech on Cities of Peace as a vision for Ottawa. I asked the crowd if their light was fierce and were they ready for tomorrow's child, not yet born. This child has difficult questions: "What did you do when there was still time to create a sustainable world?" "On your watch, was there intelligent life in humanity's leaders and decision makers?" A resounding affirmative was delivered by the diversity gathered on this day.

* Mayor Jim Watson had this to say in 2011: "Friends for Peace is an outstanding organization that does very important work, promoting, strengthening and maintaining peace, planetary care and social justice within our communities and the environment."

*On the 10th anniversary in 2012, MP Olivia Chow received a posthumous Peace Award on behalf of her late husband Jack Layton, who was the leader of the New Democratic Party. The onstage performers were outstanding, highlighted by the world premiere of "To Young Canadians." A tribute to Jack Layton performed by Orkidstra, who commissioned composer James Wright to create a song from Jack Layton's letter to the nation. They enjoyed a prolonged standing ovation.

* The Big Soul Project (fifty singers and a four-piece band), headed by Roxanne Goodman, has appeared at the Friends for Peace Day every year They were excellent in delivering such numbers as 'What are we going to leave behind when we leave?' Its message: 'When I leave this world, will I make a mark that I was here?' The implication is that what we do today will affect the quality of life tomorrow.

* The Metis storyteller Robert Lavigne titled his talk 'Idle No More' to highlight the urgency of dealing with the misdeeds of the Canadian government with aboriginal peoples.

* I presented the annual Peace Awards in 2013. The first Peace Award went to Douglas Cardinal, a visionary world master who was the architect for the Canadian Museum of Civilization. Born to a German/ Aboriginal family, Cardinal said that all of Nature including human beings are interconnected. 'Life springs through every being and rock in this life. This is a symbiotic relationship of life and nature. Each person has divine creation in them…. That is our legacy. We have the capacity to create as well as destroy….Elders trained me to honour culture as the peacemaker. We come from a society where everyone is noble, unique and responsible.'

* The second recipient of the 2013 Peace Awards was Amber Lloydlangston, historian at the Canadian War Museum and the key person who developed the Peace Exhibit there. I praised Amber for her excellence in producing such a unique exhibit, beginning with the Aboriginal Six Nations story. Amber said that she was humbled to be present with the renowned Douglas Cardinal. In the Peace Exhibit, she said that she and her colleagues wanted to show to Canadians what peace means in the form of diplomats, soldiers, peacekeepers and humanitarians.

The Friends for Peace Day became an awesome, diverse, unique Ottawa experience. It is made possible by the generosity of volunteers, supporters and the diversity of Ottawa who show up to have a good time, be educated and inspired. It creates an epicentre of intent and action, intense at times as people are moved to both tears and laughter. It is fun, poignant and direct. The intensity and joy ripples through the diversity, all generations, faiths and cultures in our northern city. The force of the epicentre roars through the community and activist tables, Muslim families, Asian groups, elders, young folk and the volunteers. The diversity of Ottawa gathers, listens, dances, laughs, cries, and takes home an unforgettable experience of hope and confidence.

Each year Friends for Peace presents Peace Awards to Canadian citizens who have devoted their lives to securing peace, planetary care and social justice. They are given to outstanding citizens such as Grandfather William Commanda, Marion Dewar, Tina Fedeski, Bruce Cockburn, June Girvan, and Elizabeth May to mention only a few. Our mandate for peace, planetary care and social justice is always solid throughout the day, at the Welcome and Community Tables, the Silent Auction, Connection Centre and Food Court. Citizens leave at the end of the day feeling uplifted, confident and connected. The intent is to create a different form of peaceful expression that appeals to a wide cross section of Canadian citizens who want to create infrastructure in our institutions that value peace and planetary processes.

To pick up on a theme introduced in Chapter 1, I have always thought of the present millennium as the century of the daughters. Not so much as a gender separate thing, but as attributes of a holistic, nurturing presence of mind. The connection with the feminine was naturally carried into my peace activism. I have been musing about this, while reflecting on the annual Friends for Peace Day in Ottawa.

When I founded Friends for Peace Canada I was making a conscious choice to focus on the local, my home city of Ottawa. My focus was on mindfulness in schools, city environment, youth at risk and on the empowerment of women. I am astonished by the results, more true to say "blown away." For at the local level there was continuity with great women who made sure good things happen.

Many of the Peace Award recipients are women. The funds raised from the day are used to issue Peace Grants to organizations, in Ottawa and internationally, and we make a point of honouring women who run organizations that make a significant difference.

In particular we have supported youth organizations to burst on to the local scene guided by kick-ass women. Orkidstra, led by Tina Fedeski, gives children from under-served communities the opportunity to learn a musical instrument and sing in a choir. It is modelled on the Il Sistema project, which was so successful in Venezuela in breaking down barriers of poverty and violence. In Ottawa, Orkidstra is creating a quiet social revolution on the backs of children – in a very healthy way.

The Dandelion Dance Company is the creation of Hannah Beach, who brought forth a dozen young lady actors, dressed in black to several Friends for Peace Days. This Ottawa based youth dance theatre company explores social issues through movement. Their repertoire is driven by the experiences, reflections and passion of young women, 12-18 years of age. The themes they dance include children's rights, hunger, authenticity, bullying, drug addiction, stereotypes and inclusiveness. Their performance of John Marsden's "Prayer for the Twenty First Century" brought the entire audience to their feet applauding their passion for nonviolence and basic rights for people. The dance alluded to our hope and dreams we want for our society. They provide the means to galvanize parents, friends and volunteers so that good kids are created and excellent citizens emerge. Peace, Planetary Care and Social Justice are alive and well in our northern city. A Circle of Nations no less.

There is now a two week Peace Festival in Ottawa that precedes the Friends for Peace Day. It has all grown in ever increasing concentric circles. The foundations of mindfulness through the fifty organizations we partner with have taken root in our northern city. All adhere to some form of the Friends for Peace mandate: peace, planetary care and social justice. I felt that these efforts could infuse global networks from the epicentre created in my home city. I had received many invitations to be a global speaker and teacher, yet realized that a concentration on my home city of Ottawa was the primary focus. I responded to the many international invitations with a gracious decline. I was inspired to devote my time and energy to moving things just a little bit in my city, so that more good things could begin to happen spontaneously. As I soon discovered, there were many good friends across the city more than happy to make this possible. I love my summer job. It is such a rewarding experience.

Prayer for the Twenty First Century – by John Marsden

May the road be free for the journey,

May it lead where it promised it would.

May the stars that gave ancient bearings

Be seen and be understood.

May every aircraft fly safely,

May every traveler be found,

May sailors in crossing the seas,

Not hear the cries of the drowned.

May gardens be wild like jungles,

May nature never be tamed.

May dangers create of us heroes,

May fears always have names.

May the mountains stand to remind us

Of what it means to be young,

May we be outlived by our daughters,

May we be outlived by our sons.

May the bombs rust away in the bunkers,

And the doomsday clock be rewound,

May the solitary scientists, working,

Remember the holes in the ground.

May the knife remain in the holder,

May the bullet stay in the gun,

May those who live in the shadows

Be seen by those in the sun.

8: Taking Refuge in Grandchildren

Taking refuge can provide delightful surprises.

It is not always a Zen teacher, wise nun or high monk who is there to provide guidance and insight. My grandson Callun has provided quite a few for me.

Callun's home is in the town of Nanaimo on Vancouver Island in British Columbia.

One summer holiday Carolyn and I spent a sea kayaking adventure with Callun and his father Iain, exploring the fascinating coastline of Vancouver Island.

On one occasion when Iain and Carolyn went shopping, I stayed at the house to meditate.

Callun was playing outside.

He came in crying after a while and tapped me on the shoulder and said: "Grand Pooh Bear (what he called me when he was a little boy), sorry to disturb your practice but I've been stung by a bee on my neck and it hurts."

I opened my eyes and took Callun into my arms and said, "My dear Callun, you *are* my practice."

I gently took the stinger out of his neck, put some ice on it and cuddled him for a while before he happily went outside again to play.

He had brought home to me that all of life is my practice. To my grandson Callun I bow in gratitude for being such a mindfulness bell for me. When I take refuge in this manner, I am aware of Buddha nature being graciously presented to me.

Another grandchild, Millie, sent me some drawings for my birthday quite a few years ago. With her five-year-old's determination she endeavored to draw a picture of me with no feet, only one arm, with a fuzzy beard, jug handle ears and much slimmer than in reality!

Over my head she had drawn a yellow halo, which is totally undeserving, yet I learned that this is how Millie thinks of me. She was revealing her Buddha nature to her grandfather and I joyfully took refuge in her love and kindness.

Several years ago, after leading a meditation retreat on the British Columbia mainland I arranged to take a ferry across to Nanaimo on Vancouver Island to visit with my son and grandson Callun. It was a beautiful calm sea voyage with the sunset dancing in the wake of the ferry. Although I was tired from the retreat, this was a delightful respite.

Both Iain and Callun were there as the boat docked in Nanaimo. As it was almost Callun's bedtime, he asked if I would read him a story once we got to their home. I was happy to do this.

Callun quickly changed into his pyjamas and chose a story for me to read. I lay down on his bed beside him and started to read. In only a few minutes I was fast asleep. My son, Iain, on hearing the silence, came into the bedroom and saw that Callun had pulled the bedcovers up over me and was sitting up in bed with one hand resting lightly on my shoulder, a beautiful smile on his face as he took care of his grandfather.

My son was moved to tears by this. He drew a chair into the bedroom and sat there with us all night. He did not want to miss the magic. Three generations taking refuge in one another. Totally present, hearts wide open. Only one snoring, but gently.

PART 3:
HEALING & TRANSFORMATION

"The first act of a newborn baby is inhalation. The last act of a dying person is exhalation. The focus of 'Death Breaths' as a transformative energy is on extending the component of exhalation so the mind becomes convinced that this is the last breath to be experienced. I was taught that this creates a different internal 'space.' It is at this juncture that energy shifts away from body survival as the mind's primary focus, being diverted to the realm of the deep unconscious. This brings about dramatic somatic and cognitive effects."

- White Eagle Woman

9: **Healing Journeys**

I first met White Eagle Woman at a gathering of elders in 1978. She called me over and asked if I was Ian? Her next words made me distinctly uncomfortable, as she announced that she did not like me at all.

I remained silent, not realizing that this blunt introduction to White Eagle Woman was the prelude to a thirty-year period of training and healing under her guidance.

She was a heavy-set woman with a round face and long black hair, but it was the air of quiet authority in her presence that forcibly struck me. She rarely smiled, but when she did, it illuminated the entire room, her dark eyes lighting up with mirth. Her rebuke was perhaps well deserved, given how unaware I must have seemed.

My disjointed education and experience with the indigenous domain of mysticism did, however, slowly evolve into a seamless pattern under her tutelage. I was very fortunate to learn from her. I discovered, much later, that I reminded her of the white man who had sexually abused her when she was a teenager. Our work together was mutually beneficial, as I believe she grew to like me, at least I hope she did.

Shamans and medicine people from far and wide came to consult with White Eagle Woman. Elders from the Amazon would come to see her. She was a holder of the Midewiwin lineage, a secret tradition of medicine people which stretched far and wide across the Americas. She was also the first of four incredible medicine people who provided me with shamanic training and teachings over the next three decades.

Their guidance enabled me to restructure my consciousness in ways that enhanced my sense of wholeness. The altered states of consciousness that I experienced loosened the control imposed by my waking mind on limiting perceptions and cognitions.

Furthermore, the shamanic healing journeys undertaken with them induced the release of the energies of deep-seated traumas. Knowledge I did not know I possessed would surface and support the process of transformation. The work with these four shamans was a highly privileged encounter.

White Eagle Woman directed the shamanic process of my healing from childhood sexual abuse, allowing the past to start revealing itself. At that first encounter in the elder's gathering she told me about a Vision Quest on her reserve in Sault Ste. Marie in southwestern Ontario. I was to be there, as she had received instructions from her ancestors to train me. That was enough to promise my attendance.

The eight day Vision Quest began and finished with an *inipi* - a sweat lodge. In between were six days of fasting, prayer and ceremony in the wilderness.

White Eagle Woman located me in a small grove of birch and oak trees and I had to stay within a strictly designated area. The other participants in the Vision Quest were placed in a different part of the forest, distant and unseen. I found level ground for my tarp and strung it over a frame built from what I could find within the grove. I placed my coloured ribbons at each of the four directions, also for the realms of above and below.

One of the oak trees became the symbolic stem of my pipe. The bowl of the pipe was a clamshell with tobacco in it. As the sun moved the tree's shadow, I had to be alert and move the clamshell in the same direction around the base of the tree.

I was very still and silent, observing my territory's nuances, the leaves, smells, insects and the rain, while in a constant state of prayer and thanksgiving. White Eagle Woman stationed herself in a trailer close by for anyone who needed guidance. She indicated that a medicine bear would visit us and to report that to her. Time passed in a seamless flow, scarcely existing before we gathered, eight days later, for the final sweat lodge of the Vision Quest.

On coming off the land, I had to consume a half cup of blueberries and then drink vast quantities of a foul tasting concoction created by White Eagle Woman. This was a cleansing medicine to make me throw up the blueberries.

It was quite disgusting. Especially for me, as it took a long time before I vomited up the blueberries. White Eagle Woman's comment to me was terse. She pursed her lips and looked at me quizzically:

"Hmmm, you're holding on tight to resist the truth you need to know!"

I had no idea what she was referring to. Once we were all in the sweat lodge, White Eagle Woman asked about the medicine bear. Nobody reported experiencing it. In exasperation she turned to me and announced she had seen the medicine bear visit me twice. What did I remember?

I recalled dreaming about a tall, gangly and somewhat goofy creature that was not a bear in my mind. I had also noticed the creature on another day, out of the corner of my eye, sitting next to the sacred oak tree.

White Eagle Woman immediately threw tobacco on the fire to absolve my ignorant gaffe and instructed me that a medicine bear can take on many forms. The goofy creature was the most receptive one to an idiot like me. Though the medicine bear had been easy on me, White Eagle Woman was certainly not.

She chastised my lack of insight while we were all in the sweat lodge. Later on, in private, she quietly revealed the door that had been opened wide for me due to the medicine bear experience. His visit was to assess whether I was capable of receiving medicine gifts from the past.

The significance of what my mentor was saying did not penetrate very deeply, although on a daily basis I followed White Eagle Woman's instructions. I would come to a stop, look deeply at my surroundings and enter a dialogue with the feminine seeds of knowledge in my consciousness. I listened in the silence to communications from the feminine wisdom within me to address issues and questions, letting it become my fieldwork of life. I made diary entries with my questions and dilemmas, doubts and misgivings and waited quietly for a reply from within. I wrote with respect, love and gratitude and anticipated counsel to arise from inside. It was frequently not what I expected, hence my faith in its integrity.

White Eagle Woman also ensured that I trained with other shamans in journeying, so I could eventually create a healing journey for non-native people. She trained me in a breathing cycle for use in a Healing Journey.

The breathing cycle begins with deep breathing; In-breath: Holding Fullness: Out-breath: Holding Emptiness. There is a slow count for the Holding Fullness and Holding Emptiness portions of the deep breathing. The count increases to holding the fullness and the emptiness for slow counts of twelve. This build up took approximately twenty minutes. It was then followed by several minutes of short explosive breaths, after which the death breaths begin. These are long in-breaths and retention of breath for a slow count of twelve, followed by the out-breath and holding the emptiness for a slow count of thirty. The holding emptiness continues to expand to building up to holding the emptiness of breath, after the exhalation, for as long as possible.

White Eagle Woman told me that the first act of a newborn baby is inhalation. The last act of a dying person is exhalation. The focus of 'Death Breaths' as a transformative energy is on extending the component of exhalation so the mind becomes convinced that this is the last breath to be experienced. I was taught that this creates a different internal 'space.' It is at this juncture that energy shifts away from body survival as the mind's primary focus, being diverted to the realm of the deep unconscious. This brings about dramatic somatic and cognitive effects.

Death breaths were the key part of a cycle of breathing that I was trained in by my shaman teachers. My experience of this cycle, cognitively and physiologically, varied in the many shamanic journeys undertaken, but there were common experiences. Firstly, the deep breathing took me into a state of mental calm. Secondly, the short explosive breaths produced a sort of portal or gate that I could feel myself going through. I also felt light headed and dizzy.

The death breaths were very different. On the in-breath and retention I would feel stable, on a particular plateau of experience. On the out-breath and holding emptiness for as long as I could, my limbs and body would shake and tremble. There would also be periods of profuse sweating and extreme cold, particularly in my hands and feet. Then on the last gasp of holding emptiness I would take an in-breath. The trembling and shaking would stop, and while holding the fullness of in-breath for a count of twelve I would experience a deep clarity.

Cognitively, I felt hyper alert and hyper lucid. The next out-breath and the holding of emptiness for as long as I was able, would again produce trembling and shaking throughout the body. I felt extreme heat and cold but from a different level of experience. Once more on the last gasp of holding emptiness after out-breath, the in-breath produced a feeling of cognitive calm but at a totally different level.

In doing the death breaths over a period of time, I felt I was taking steps to different plateaus of cognition. In the process, I always felt a sense of control, totally cognizant of what was taking place at different levels of awareness.

In addition to breath, the other drivers experienced with my shaman teachers were music with a repetitive rhythm, drumming, rattles, and nature-based music that incorporated the sound of animal calls.

Each cycle of deep breathing, explosive breaths and death breaths would be followed by a period of normal breathing for several minutes. Then the entire breathing cycle would be repeated with a different driver. I would experience similar cognitive and physiological shifts, once more advancing to other states of consciousness. I was aware of being in an altered state, yet simultaneously existing in conscious reality.

With the death breaths, I could sense a different and expansive "space" opening up in my consciousness. My mind was being convinced that this was the last breath my body would experience. In actual fact it was not, but the trick was to move my mind into the *perception* that it was. The energy devoted to physical survival could then be diverted into triggering a connection to unconscious material that I could dialogue with.

Then one is considered to be ready for the shamanic journey, which involves symbolic imagery while you are in an altered state of consciousness. It provides a loose structure so the individual can dialogue with archetypal material about their unique personal scars, traumas and also about their creative potential. The dialogue experienced in the journey can continue afterwards through writing in a journal, communicating the experience and consulting with the shaman facilitator. At the end of the shamanic journey you return to where you are sitting or lying down.

From the experiences with White Eagle Woman and shamans in Arizona and Ontario, I carefully put together a Healing Journey suitable for non- indigenous people. This was White Eagle Woman's intent in teaching me, so her knowledge could be passed to a wider audience.

The healing ceremony is based on principles of safety, sacredness and responsibility. It draws on two traditions, shamanism and Buddhism, and my extensive experience of both. Sacred Native American flute music is played quietly as people gather for the ceremony, form a circle and participate in the opening purification of smudging.

The healing journey begins with a stated focus on conscious breathing, so that everybody's awareness becomes attached to in-breath and to out-breath. Individuals gather in a circle and burning sage is smudged over each person in a clockwise direction.

During this opening process, a musical driver is in operation. I always emphasized that people validate their experiences from their own internal recognitions. This can be discussed and verified at the end of the ceremony.

The next step, a symbol of internal purification, is a simple heart center meditation. During this meditation the sound driver changes from the opening sacred flute music to the sound of the eternal OM, Gregorian or Blessing Way chants.

Here are the instructions: As you fill your lungs with the in-breath, visualize white light coming into the middle of your chest: your heart center. You can visualize this as light floating gently down to the area behind the sternum, or as a funnel of light coming directly into your chest from the universe. If you do not visualize easily, then *think* the light coming into your heart center.

Feel this white light as a gentle glow and take it through to your heart center, inside the chest and throat, up to your crown. All this is on the in-breath. At the end of the in-breath, at the top of the crown, hold the breath for a moment and place thought into the procedure. The thought is: 'send this light to every cell in my body.'

Then, on the out-breath, imagine the white light moving from your crown, filling your entire body right down to the toes. You complete the process by grounding the energy through your feet into the earth. Do this ten times or until you feel a sensation or feeling of relaxation. Remain within the silence and energy of the meditation for approximately ten minutes. If thoughts distract you from the process, simply come back to the focus and direction of breath, light and word. This meditation is a centering vehicle that grounds the person in their body and is an important step to ensure stability.

The next stage introduces two drivers, used in combination. The first breathing cycle of deep breaths- explosive breaths- death breaths begins. It is accompanied by a different sound to that of the prior meditation. Changes in tone and rhythm intensify the driving effect of the breathing cycle. The initial breathing cycle of deep breaths- explosive breaths- death breaths is to the accompaniment of a tonal musical driver. I use new age synthesizer music or chants that are repetitive, relaxing and harmonic. After the first breathing cycle is finished, everybody relaxes their breathing during the pause and prepares for the second breathing cycle. This second cycle is accompanied by music that has a more insistent, repetitive, driving beat along the lines of the electronic synthesizer music from *Chariots of Fire*. Once the second cycle of deep breathing- explosive breaths- death breaths is complete, the third cycle begins. The sound driver changes to ten minutes of repetitive drumming.

Each breathing cycle is associated with distinct sound drivers. Once the last breathing cycle is complete, normal breathing ensues. The individual by now should be in an altered state, extremely relaxed. It is at this point that the shamanic journey begins with extended symbolic imagery and a different sound driver. The healing journey is accompanied by softly played nature-based music that incorporates animal and birdcalls, and other sounds drawn from the world of nature. I use Dan Gibson's "Algonquin Suite" as an appropriate choice for the healing journey.

As I discovered in training, there are many variations to a shamanic journey; into the past, the future, under the sea, into the earth and beyond time and space. They can be guided or non-guided. The initial preparation grounds the person taking the journey in their body and in the present moment. This is essential so that past fears and future anxieties that might arise during the shamanic journey can be dealt with from a solid foundation. The following is only one form of guided symbolic imagery. The "Algonquin Suite" would play softly in the background.

See yourself walking through a beautiful meadow, full of flowers. You hear the sounds of insects humming, and birds singing. The sun feels warm on your face and a slight breeze ruffles your hair. Look up into an endlessly clear blue sky and for a moment allow yourself to merge with it, and enter such clarity. (*Pause*)

Notice a small shape hovering in the sky that gets bigger as it comes closer to you. It is a golden eagle slowly circling above you. He is your guardian and will watch over you and keep you safe on your journey. As you walk, the meadow slowly gives way to a river that runs over rocks before eddying into deep, still pools. Follow the bank of the river in the direction of the sun. There is a path to walk along. Notice the mallard ducks at the water's edge with their ducklings, and a kingfisher sitting patiently on a

113

branch overhanging a deep, still pool. The sun filters through the trees at the river's edge and the light dances on the rocks and water like a crystal cloak that shimmers and moves with every swirl and eddy. **(Pause)**

Walking around a bend, you see that the river runs from a clear lake fringed with forests, reflecting snow-capped mountains in its still surface. Find a spot by the side of the lake, sit down and enjoy the intimacy of nature that is around you. At the end of the lake you see a cow moose with her calf at the water's edge. In the distance you hear wolves calling to one another and notice two rabbits beside a shrub close by. A doe and two fawns walk slowly and tentatively from the forest into the sunlight. Skylarks hover motionless in the sky then descend to earth with their lilting song. Your eyes are drawn to a stately blue heron standing motionless in the reeds at the lake's edge. These creatures are there to remind you of your connection to the world of nature. Take a moment to be with the grass, the trees, animals, birds, insects, and bring to this place your favourite animals. **(Pause)**

Ask one of the creatures to accompany you on your journey and wait to see which one comes forward. It does not matter if none come forward, the golden eagle still circles overhead as your guardian. **(Pause)**

After sitting by the lake's edge for a while, stand up and slowly walk into the water. It is icy cold, fed by glaciers from the snow-capped mountains. But it is a cold that is bearable because it purifies, stripping you of your anxieties, stress and worries. Slowly walk into the water up to your hips, your chest and then submerge yourself in the icy cold embrace of purification. Underwater, you can breathe and move around with ease. Notice the rays of sunlight entering the water, fish swimming swiftly past and see the rocks and submerged tree trunks on the lake floor. As you move around and adjust to the water you see a cave at the bottom of the *lake and you swim strongly and powerfully to enter it. There is light at the end of a long underwater passage and you swim through and emerge out of the water into a cavern covered in crystals. The sound from the crystals shimmers through your body. At the edge of the cavern is a waterfall. Stand underneath it and feel the water washing over and right through your body. Feel the energy of the waterfall taking away your anxiety, tension and distress. **(Pause)***

*Leave the cavern and follow a trail that takes you through a pine forest. Beautiful tall pines are on either side of you, stretching to the sky. Take a moment and see the blue sky endlessly clear and enter such clarity. **(Pause)***

*Then see the forest open into a large clearing with a big flat rock in the center. There is a fire prepared for you by the rock. As you warm your hands by the fire and feel its warmth on your face, there is a presence next to you. Turning around, you see a beautiful old woman with clear brown eyes that look right into you. She smiles in welcome and you feel she knows all about you and embraces you in a simple, heartfelt love. She is a very powerful healer, a wise shaman and is there on your journey to serve you. **(Pause)***

*Standing next to her is a handsome old man, with weathered features and a gentle smile that lights you up. From his eyes, you feel overwhelming compassion and understanding. He is a very powerful healer and a wise shaman and is there on your journey to serve you. **(Pause)***

Between the old man and old woman is a young woman who sparkles. She is fresh, vibrant and beautiful, aglow with life's vitality. She also greets you with a smile and a love that you know is unconditional. She is the feminine source of Earth Wisdom and a lightning rod for your transformation. She knows very well the suffering and chaos of modern times. She is a very powerful healer and a wise shaman and is there on your journey to serve you.

*Particularly if you are at the crossroads of New Beginnings and ready to discard the old damaging tapes you run in your mind. Her power has an infinite depth and force. (**Pause**)*

*Know that these three shamans also come from the deepest part of yourself and they represent your own powers of creativity and self-healing. The three shamans approach you and invite you to speak to them. Choose who you wish to communicate with, and talk about whatever distresses you: the anxieties of the day, the stresses at work and at home, then if you wish, go deeper into your distress. Talk to them about growing up, the neglect and abuse you may have experienced, the isolation, separation and lack of understanding you encountered as a young person, adolescent and adult. Talk about the damage caused to you and the damage you have caused to others. Talk about the hatreds, angers and insensitivities you experience and perpetuate. You can say anything to these three shamans. They understand and love you and are there to heal you. Talk about whatever you feel free to communicate and feel the distress and trauma leaving your body. And when you run out of things to say, just be with their loving and supportive presence. For now, open up and speak to one of these immensely powerful shamans placed on your path. (**Long Pause**)*

*Ask each one of them if they would transfer their power of creativity, understanding and healing to your awareness. Of course they will agree. Look into the eyes of each one of them in turn and feel the transfer of their healing power with a jolt or energy circulation within your body. Thank them for this gift, then ask if you could speak to someone from the other side. Someone who has passed on, that you did not have the opportunity to share what you wanted to say, or hear what you would have liked to hear. Wait and see if anyone comes and do not be disappointed if nothing happens. It is not the time. (**Pause**)*

Take your leave of the shamans. Thank them for their support, love and power of healing. Turning around, you see a beautiful child surrounded with a golden aura. This golden child is you, without trauma, wounds or damage. The child comes directly to you and takes your hand, and leads you to a cliff edge where the beautiful golden eagle is waiting for you. He has been there as a guardian throughout your journey and is now ready to take you home. **(Pause)**

Ask your golden child if he or she wants to come with you, then climb onto the back of the eagle, and feel him take off from the ledge and soar high on the updrafts. Below you, see the mountains, lakes and forests of your journey. Smoke curls lazily skyward from the fire by the rock and as you fly with the eagle feel how beautiful this earth is. Then when you feel ready, leave the eagle and fly on your own with your golden child next to you. With your arms spread wide as wings, catch the air currents and soar, then swoop low over the streams and mountains and enjoy the strength of flying on your own as your golden child merges with you, creating one unified being. **(Pause)**

Slowly fly back to the edge of the lake where you were sitting. Once again notice the animals, birds and insects and see how happy they are to see you again. Sit there for a time. **(Pause)**

See yourself lying down in the healing circle. Form a circle of brilliant white light around where you are lying, then step through the light and slowly return to your body. Breathe deeply on the in-breath and exhale deeply on the out-breath. As you breathe in, say quietly to yourself, "I have arrived." As you breathe out, say quietly to yourself, "I am home." Continue to do this breathing exercise for at least five minutes or until you feel "arrived" and "home" in your body.

After the safe return, a final meditation with light is conducted in a circle.

A tray of lighted candles is passed around in a clockwise direction and each person acknowledges the light and healing in the next person in the circle.

The entire healing ceremony has been about surfacing, clearing fetters, knots and blockages, in other words releasing energy "sinks."

The internal dialogue with the shamans at the rock is with the powerful archetypal material of creative self-healing that exists in everyone.

The final meditation with light acknowledges that there is more to consciousness than trauma, suffering, blockages and energy "sinks."

There are seeds of happiness, joy and grace that acknowledge the inherent divinity within everyone.

The acknowledgement in the final meditation nourishes these seeds in consciousness. This creates a crucial finale to the healing journey.

This healing ceremony synthesizes my training and personal experiences in meditative and shamanic practices.

It is based on a model of healing which assumes that blockages in the mind/body/inner nature system are caused by traumas from conditioning, genetic heritage and karma.

These traumas lodge themselves, as energy "sinks," in the body and mind.

The removal of such blockages must also be in energy terms in order that a higher level of integration can take place.

Throughout the breathing cycles and journey, other material from the depths of unconsciousness will surface.

118

It is essential to be aware and dialogue with all of it, so that the energy of trauma is steadily diminished through mindfulness and the power of self-healing.

The initial focus on meditation and breath work secures the individual within the safety of his or her own body, quiets the mind, and provides a foundation for the knots of suffering to surface during the breathing cycles and the journey.

The surfacing can be somatic, mental and emotional, particularly so during the dialogue with the three shamans at the rock.

A major significance of the dialogue process is that each individual chooses which level of distress to communicate.

No one is pushed to deal with more than they can cope with. The ceremony clearly emphasizes the therapeutic role of an altered state.

The death breaths focus on the extension of exhalation so that "space" is created within the individual's consciousness for distress to more readily surface and be transformed.

The release of distress is tied to the imagery of the journey and the psycho-physiological triggers induced by the breathing cycle.

The intent and focus of the entire ceremony is to heighten the significance of the dialogue by creating a different dimension for consciousness to facilitate our healing, well-being and integration with ourselves.

"Emotional, physical and sexual abuse during childhood can create a lost, frightened and frozen child within us. If we are unable to reach this lost and wounded child, then we may never heal ourselves. We prefer not to remember the sufferings of childhood, so we bury them and hide. We run away from seeing deeply into the causes of our suffering. Whenever the memories arise, however fleetingly, we think we cannot handle them and deflect them into the deepest realms of our unconsciousness mind. This results in the wounded inner child not being seen for a long time, simply because we are terrified of further suffering. Yet we have to find a way to reach the hurt child and make them safe. This means we have to get past the fear and address the suffering, realizing that it is suffering which provides the way through to awakening."

- Dr. Ian Prattis

10: Healing the Inner Child

The Territory of Suffering

"Dealing with suffering is like handling a poisonous snake. We have to learn about the snake, and we ourselves have to grow stronger and more stable in order to handle it without hurting ourselves. At the end of this process, we will be ready to confront the snake. If we never confront it, one day it will surprise us and we will die of a snake bite. The pain we carry in the deep levels of our consciousness is similar. When it grows big and confronts us, there's nothing we can do if we haven't practiced becoming strong and stable in mindfulness. We should only invite our suffering up when we're ready. Then, when it comes we can handle it. To transform our suffering, we don't struggle with it or try to get rid of it. We simply bathe it in the light of our mindfulness."

I employ the above wise words from my teacher Thich Nhat Hanh (*Reconciliation: 2010, 64*) to open this chapter on Healing the Inner Child. First of all, we have to develop and nurture our mindfulness, which means understanding the reality of our suffering that we would rather avoid. There are clear warning signals, if we choose to pay attention, but we get caught in our dramas and telling our sad stories to whomever will listen. We also court our suffering, keeping it alive, often engaging in a competitive sport: my suffering is bigger than yours. The courtship of suffering can be an ugly romance for we enter into a co-dependent relationship between our mind and suffering.

Physiologically and emotionally, we become so tightly tied into our suffering that we cannot be without it, even though it is destroying our well-being. We grasp at brief insights into our suffering but deal only with surface appearances. Yet the surface exposure has a long history of gathering momentum until it actually surfaces. The small snake has become a monster. The addiction to suffering is now embedded in our mental state.

We respond to any glimpse of suffering with such destructive emotion that we reinforce the causes and conditions that created the suffering in the first place. And so we continue shooting ourselves in the foot, torpedoing our lives over and over again.

Our suffering is largely caused by abuse: emotional, physical and sexual, and it becomes an organizing template in our mind. We can create an abusive relationship with that template's qualities: addiction, fear and co-dependency. To stop the cycle of harm we need a moment of realization. 'This is what I have been doing all my life? How do I stop it? That insight has to arrive in the mind before we can apply ourselves to developing mindfulness as an antidote to our suffering.

It is amazing to penetrate the darkness and realize that the abuse you have suffered has created a negative relationship with yourself. Mindful engagement can bring the abusive relationship to a halt. This is the required moment that propels you to get to work, to investigate the causes and conditions that placed you in such suffering. We learn the practices, tools and concentrations that support this journey to understand suffering and take care of it. We break the cycle through re-training and mindfulness practice. We equip ourselves for a journey to be well that requires our determination to practice mindfulness daily and ensure that we take refuge in wise support.

The Wounded Inner Child

Emotional, physical and sexual abuse during childhood can create a lost, frightened, frozen child within us. If we are unable to reach this lost, wounded child, we may never heal ourselves. We prefer not to remember the sufferings of childhood, so we bury them and hide. We run away from seeing deeply into the causes of our suffering. Whenever the memories arise, however fleetingly, we think we cannot handle them and deflect them into the deepest realms of our unconsciousness mind. This results in the wounded inner child not being seen for a long time, simply because we are terrified of further suffering. Yet we have to find a way to reach the hurt child and make them safe. This means we have to get past the fear and address the suffering, realizing it is suffering that provides the way through to awakening.

Although we're now adults, there's still a child inside, afraid and suffering deeply, regardless of any happy face we present. This hurt child inside colours everything we do, generating our fears, insecurities and self-loathing, wounding us in our relationships and life. We must extend a different energy to them so that childhood suffering is understood, defused and transformed. Mindfulness is the way through to the inner child. We have to skillfully embrace them exactly where they are caught by the past, in fear and with anger at being neglected for so long.

It means touching the seeds of childhood suffering from an adult state of being, mindful and aware, knowing we must make it safe for that child to come out. We, as adults, must no longer run away. We need to have the courage and awareness to bring healing to our hurt inner child and produce a transformation for ourselves. The steps we take are not only to heal ourselves. We somehow connect to all wounded children, those who were our ancestors and those who will be our descendants.

For once we cultivate the seeds of mindful healing in ourselves, the energy of these seeds continues into all that we connect with. This is a quantum leap from our cellular memories to everyone else's throughout time and space. With awareness, we take our inner child into our daily life, patiently realizing that we are on a splendid adventure to bring the cycle of suffering to a close, even if it has persisted over generations. We are healing and transforming generations of ingrained patterns transmitted from our ancestors and continuing, through us, to our descendants.

Thich Nhat Hanh addressed the issue of child abuse in a question and answer session held in the Lower Hamlet of Plum Village, France on the 17th of October, 1998. Very gently, he spoke about the ignorance and pain of the abuser as well as that of the abused, and stated clearly that understanding was the basis of recovery. Not blaming or feeling guilt and shame, but seeing deeply and understanding. First of all, the knowledge that the abuser must have lived under ignorant and deprived conditions without support, guidance or a wise teacher. The power of ignorance was stronger than the person, and they were driven to do wrong things. If the person abused can begin to understand just a little bit of that, then their anger, shame and outrage can transform into a droplet of compassion. Through practicing mindfulness, their suffering can diminish. When forgiveness and understanding are there, suffering decreases. The second step he suggested was to recommend that the abused person practice mindfulness, to transform themselves into a Bodhisattva and engender the compassion to help and be of service to all children who need protection. By merit of understanding the experience and a recovery from abuse, such a person can practice and use their talents to promote measures to protect children. This helps to eradicate the ignorance that generates abuse.

There are many techniques and methodologies of therapy that address issues of the inner wounded child. The first one I am going to describe is accessible to most people. It is a first step and I recommend that it be practiced under the guidance of a therapist, shaman or spiritual teacher. The individual starts a diary or log book for the person and their inner child to write to one another. The adult writes as normal, beginning by saying hello to their inner child. They explain how sorry they are for having been away and neglectful. Grown up now and strong, they are going to do everything to make it safe for the child, creating an environment in which they will be loved and cherished.

Using the hand not normally written with, allow the inner child to express itself. Do not edit. Just write down whatever comes out. It may be angry, blame filled and abusive content, but it is important not to be shocked or defensive, providing constant re-assurance, love and guidance. The adult inside must bring the wounded inner child qualities of love, compassion and wisdom. These are the seeds of mindfulness to support the wounded child inside. The energy of these seeds works on the energy of the traumatized inner child to reduce their pain and suffering. Talk to them through writing with total love and acute mindfulness. Read the diary entries out loud, in both the roles of adult and inner child. This simple act of reading out loud is a way for both aspects of the person to be heard. On a daily basis, register with how deeply understanding and love is getting through to the wounded child, for they are listening carefully to every word and know that the adult is now listening. As the adult and inner child grow closer, awareness, love and healing is brought to the suffering and pain of the child.

Details of trauma may be revealed, which is why the help and guidance of a trusted therapist, shaman or spiritual teacher is essential. It supports the wise and loving parent relationship with the wounded inner child. With time, there will be shifts and changes in patterns of expression as the child becomes trusting and starts to grow, eventually merging fully with the adult. In letters talk about yourself and your life, take your inner child on outings, giving them treats, care, attention and love that may have been lacking in real childhood. The suffering will diminish and will transform relationships with co-workers, friends and family. Fears of the past and anxieties about the future will not have the same driving force. Noticing the results, thank the inner child: "Thank you for being with me. That makes me so happy." The experience of being with the inner child in the healing journey is a stimulus for this kind of happiness. This process can create a wide range of strong emotions, which is why guidance and support is necessary on beginning the journey, remaining steady and mindful.

I know, for I went through it. I am happy to say that it worked for me, as I experienced the painfully slow establishment of trust, then the exhilarating joy of safety and integration, until finally my inner child was the adult me, integrated with a freshness and vitality that I continually treasure.

To support this journey, there are other practices and meditations that are valuable for the steady process of healing. Buddhist teachings contain a multitude of tools, concentrations and practices that can nurture this process. They include: The Five Remembrances, Five-Year-Old Child Meditation, Sutra on Mindful Breathing, Deep Relaxation, Touching the Earth and Removing the Object. In Plum Village, Thich Nhat Hanh's practice center in France, he has provided a much-loved practice *gatha* for the meditation community, which begins with "I have arrived, I am home."

This is used in walking and other meditations as an instrument to concentrate on breath and be present. In this way, the fears and traumas of the past and anxieties about the future do not crowd in and overwhelm the mind.

The *gatha* with walking meditation, connected to in-breath and out-breath, provides an essential tool to take care of the many mental formations that flood our waking consciousness with fear, pain and suffering. With daily diligent practice, we can examine the same mental formations, but from a place centered in mindfulness. This simple *gatha* has become the dharma seal of Plum Village.

Inner Child Has Arrived Meditation

The Vietnamese origin of the *gatha* provides a penetrating tool to touch our inner child, who suffers from trauma and abuse experienced in childhood. It does not translate as "I have arrived, I am home." It translates as "Your child has arrived, your child is home." This is so beautiful to say to yourself as you breathe in and out whenever you do walking meditation, for each step welcomes your wounded child to be well and to come home to you. When you walk to your car or your office, by a river or in a park, you can be specific and recite to yourself:

In-breath   "My inner child has arrived"

Out-breath   "My inner child is home."

This is good practice, for, with intelligence, you welcome your wounded inner child home through the practice of being present. We are capable of arriving in every moment of practice, whether it is sitting meditation, walking meditation, having a mindful meal, taking a shower or doing laundry. Being present in each moment is a way of practice that welcomes home the injured, frightened inner child harmed by abuse.

In order to heal, it is necessary to cultivate the internal energy of mindfulness before stopping and looking deeply into what caused the fears and traumas of abuse. The practice of arriving in each moment nurtures that strength. From the space of clarity provided by locating yourself in the present moment, not only is your inner child welcomed home, there is also the lucidity of mindfulness practice to deal with the ghosts of the past and at the same time put future anxieties to rest.

Love Meditation for the Inner Child

Another tool is to adapt the Four Brahmaviharas meditation to focus on the injured inner child and is based on the Buddha's teachings on love. Prepare for meditation by sitting comfortably with the spine erect. Bring your concentration and focus on the in-breath and out-breath. After 10-20 breaths, whenever you feel calm and stable, begin bringing each of the components: love, compassion, joy and equanimity into yourself. The next sequence provides focus and concentration to water the seeds of love, compassion, joy and equanimity within your inner child.

In-breath	"I bring love."
Out-breath	"to my inner child."

You can give a loving name to your inner child if you wish. Feel the energy of love fill you head to toe and register the energy for a few breaths. Then continue the same way with:

In-breath	"I bring compassion"
Out-breath	"to my inner child."
In-breath	"I bring joy"
Out-breath	"to my inner child."
In-breath	"I bring equanimity"
Out-breath	"to my inner child."

Conclude the meditation by once more bringing love, compassion, joy and equanimity to the adult.

This meditation nurtures the wounded inner child and, at the same time, the adult.

The Buddha's teachings on love provide the foundation for this meditation to the wounded inner child. The concentration on these four qualities is an incredibly powerful instrument for healing.

There are many other methods of meditation and practices. There were practices that accompanied the shamanic healing conducted in an altered state of consciousness (See Chapter 9: Healing Journeys).

One very important factor was my determination to heal, once I understood what had happened. From that awareness, I took specific steps and relied on wise teachers, medicine women and steady friends to help me along the path of healing and transformation.

I must emphasize that this is not a journey that can be taken alone, so ensure that you have support from your community and good guidance from a therapist, shaman or spiritual teacher.

I am reminded, with gratitude, of a particular closing chant sung after ceremonies and sutra readings in Plum Village:

Practicing the way of awareness

Gives rise to benefits without limit.

We vow to share the fruits with all beings.

We vow to give tribute to parents, teachers.

Friends and numerous beings

Who give guidance and support along the path.

"My child, you have met the Buddha at the Gate. He is asking you to look deeply into the depths of your reactions and anger. He is asking you to listen instead to the deep source of love and compassion in your heart. He is asking you not to lose your joy and equanimity. He encourages you to develop your equanimity so it is solid and strong, not easily moved. These are the Buddha's teachings on love and you must meditate deeply on these teachings."

- The Abbot on Buddha's teachings

11: The Buddha at the Gate

Let me tell you a story. There was a young monk who was sent by his Abbot to beg for food in a nearby town. The town had a wall around it, with a main gate placed at each cardinal direction. The young monk was a little nervous during his first alms round but the townspeople were very generous and quickly filled his bowl. Late that morning, he decided to leave by the North Gate. Sitting to one side of the gate was a bedraggled, dirty, old beggar who stirred himself at the sight of the young monk and started to spit and curse at him. The monk jumped to one side in alarm and quickly passed through the gate as fast as he could. As he walked away he could still hear the beggar's curses ringing in his ears.

On the next day, once his bowl was full he decided to leave by the West Gate to avoid the dreadful old beggar. But the beggar was there, spitting and cursing at him once again. The young monk was angry this time and shouted at the old beggar "Don't you know who I am? I am a student of the Buddha!" At which point, the beggar picked up some dirt and threw it into the bowl, spoiling the monk's collection of food. Angrily the young monk walked back to the monastery, knowing he would have to endure an enforced fast, wondering why he should be treated in this way. So, he made up his mind to breathe and calm himself and totally ignore the beggar if they should meet again.

As he left by the South Gate next day, he met the old beggar, still cursing and spitting at him. He protected his food with part of his robe and kept his head down as he endured the abuse from the old beggar once more. His heart was in turmoil, his mind in so much distress that he could eat nothing from his bowl once he reached the monastery. On the next day he left by the East Gate and to his dismay the same old beggar was waiting for him. As he heard the curses and endured the spitting, the young monk raised his walking staff to strike the old beggar, who cackled in glee at the young monk's discomfort.

With a moment's pause the monk stayed his hand and walked quickly through the East Gate.

He was deeply ashamed at how close he had come to violence. He felt he was a wretched student of the Buddha and confused as to why all this abuse was happening to him. He suffered so much from the anger and violence inside himself that he knew he needed his Master's guidance. He sought out the Abbot and asked for forgiveness and guidance after he told the story of his past four days. The Abbot listened deeply to the young monk then smiled very gently with understanding.

> "My child, you have met the Buddha at the Gate. He is asking you to look deeply into the depths of your reactions and anger. He is asking you to listen instead to the deep source of love and compassion in your heart. He is asking you not to lose your joy and equanimity. He encourages you to develop your equanimity so it is solid and strong, not easily moved. These are the Buddha's teachings on love and you must meditate deeply on these teachings."

The Abbot instructed him on the Buddha's teachings on love, compassion, joy and equanimity; the Four Immeasurable Minds. Also known as the Four Brahmaviharas, these teachings were first given by the Buddha to a Hindu gentleman who wished to find the way to be with Brahma, the Universal God. The young monk was instructed to deepen his practice, to listen deeply to his heart and always to stop and look deeply into the causes and conditions of his reactions, anger and violence. The young monk bowed in gratitude to his Abbot and diligently practiced meditating on the Buddha's teachings, immediately putting them into daily practice. This enabled him to pass by the beggar without reaction, until one day no beggar was to be found at any of the four gates.

This simple teaching is something we can all put into practice, learning not to activate the demons in our own minds. A better world is the end result.

12: Shattering of Concepts

Huddled on a bed in an ashram in Mumbai, India, I opened my eyes to see a visiting *Swami* sitting beside me. The small ashram was reserved for saints and holy men. I did not qualify for either category, but felt their grace close at hand. One tangible and humorous manifestation of that grace was this visiting *Swami* beside my bed. He smiled broadly and helped me to sit up, then surprised me with his words:

"We are so happy, Ian, that you have decided to die with us in India. And we will be most happy should you live."

He beamed love and understanding to me. My reply, as best I remember, was to smile back and just say, "Me too!" The *Swami* made me some tea with herbs, provided a blessing and then left. When I went to sleep that night, I felt very calm about letting go of my bodily existence. I knew that the experiences of joy and freedom flooding through me were dissolving my many mistakes and bodily pain. I felt truly like me, very peaceful, no longer a maverick standing alone. Lying close to death, the lack of fear provided a sense of freedom and strength.

I had been invited for guru training in India by Rishi Prabhakar, after meeting with him several times in Canada. He recognized something in me that I did not. This was new territory for me.

I traveled to India in 1996 to teach and train in *Siddha Samadhi Yoga*. This *Vedic* tradition was ecumenical in character, a wisdom tradition totally relevant to the modern day.

By November of 1996, I had become seriously ill in India. As I observed my bodily systems shutting down, I knew there was a distinct possibility of death. To this day, I am still amazed by my calmness and lack of fear.

While in India, I was privileged to have many treasures of wisdom made available to me. There were two specific circumstances. One was Thich Nhat Hanh's book of meditations, *The Blooming of a Lotus.* Before leaving for India in 1996, at the last moment I picked up this book and placed it in my backpack. My companion during my illness and passage with death was Master Hanh's book of meditations. I was astonished by my calmness and hope to find a similar equanimity for death's next visit.

In my family and culture, there is very little discussion or clarity about death and dying, though as a child I had an intuitive understanding.

I remember when my grandfather died even though I was a small boy. I felt him as a tangible presence, even when he was in his coffin and quietly whispered to him: "Go to Heaven now, Grandpa."

I also remember at his wake, how upset I became by my relatives drinking, arguing and being disrespectful to one another. In tears, I sought out my grandmother and complained that everyone was making it hard for my grandpa to go to Heaven. She wiped my tears away with her handkerchief and listened carefully to me before walking into the living room of her house. With quiet authority, she asked everyone to be quiet and to go home.

It was much later in life, once I was exposed to Buddhist teachings on death and dying, that I realized I was not such a crazy kid after all. I had cared for my grandfather's consciousness after his physical death.

From that turning point, I knew that preparation for death was also training for life, though I did not always pay close attention to this insight.

The opportunity for liberation at the time of death was an intriguing notion. I could see that my obstacles of ego and habitual patterns of behavior were in the way of a sound preparation. I did want to merge my consciousness at the time of death with what the Sufis call "The Great Magnificence." Or if I got confused or fearful, to be able to receive guidance to do so.

From my understanding of the Tibetan *bardos,* I felt that if my death was an aware one, then in the *bardo* of "becoming," my consciousness would take a form that would serve all sentient beings. That struck a recycling chord, appealing to the ecologist in me.

The retraining of my mind was done fitfully, not in a consistent manner, until just before I left for India to take up the life of a yogi. There, the preparation became a daily practice of being aware of universal consciousness that was prepared to merge with my pitifully weak and not-so-awakened mind.

My leap of faith was that the understandings about death and dying are all in the mind. This meant that in everyday living, I could use my mind to take steps to prepare for that final moment of merging with the wisdom mind of the universe. Perhaps I would be able to do this while I was alive – the whole point.

Still, I was surprised by my lack of panic in the face of death. As that December drew towards its close, I totally surrendered. I will always remember Saturday, December 21, 1996. On that day, I let go of all attachments to my body and surrendered to a sense of freedom never before experienced.

Throughout the day and evening, I read *The Blooming of a Lotus* from cover to cover, practicing meditations that spoke to me.

I felt at one with my spiritual ancestors. I felt Thich Nhat Hanh's wisdom, love and gentleness as a tangible presence watching over me.

The meditations in *The Blooming of a Lotus* carried me over many thresholds, some of which I was aware of at the time. I could not discern the significance of most of them until much later.

The meditations took me deep into my roots of being and I felt very calm about the impermanence of bodily existence.

My heart opened wide and I thought about my many mistakes and chose not to deny them or brush aside the bodily pain. I knew that the experiences of joy and freedom flooding through me were dissolving both.

During this period of time I felt very simple, that I was living properly. I was without panic, present with whatever was happening or arising.

I did not fear death.

This lack of fear gave me a sense of freedom and strength. It opened a door to send love and joy to all. I felt truly like me, very peaceful, not pulled in any direction.

Despite all that was going on, I was within each second of time in a totally timeless way.

Whatever gifts, skills and energies I could contribute to bring joy and love to others were there to freely share. That is the only manner in which I can describe what was happening.

I finally understood the significance of the Buddha's Five Remembrances Meditation:

1. Knowing I will get old, I breathe in.	Getting old
Knowing I cannot escape old age, I breathe out.	No escape
2. Knowing I will get sick, I breathe in.	Getting sick
Knowing I cannot escape sickness, I breathe out.	No escape
3. Knowing I will die, I breathe in.	Dying
Knowing I cannot escape death, I breathe out.	No escape
4. Knowing that one day I will lose all I hold dear today, I breathe in.	Losing what I hold dear
Knowing I cannot escape losing all I hold dear today, I breathe out.	No escape
5. Knowing that my actions are my only belongings, I breathe in.	Actions true belongings
Knowing that I cannot escape the consequences of my actions, I breathe out.	No escape from consequences
6. Determined to live my days mindfully in the present moment, I breathe in.	Living mindfully
Experiencing the joy and the benefit of living mindfully, I breathe out.	Living mindfully
7. Offering joy and love each day to my loved ones, I breathe in.	Offering love
Easing the pain and suffering of my loved ones, I breathe out.	Easing suffering

By looking into these major fears, I personally experienced all of them. It made sense and carried me into a state of non-fear. There was nothing overlooked or pushed to one side. My mind was very clear. The Five Remembrances were not located in the depths of my consciousness, they were my existential reality. I neither welcomed them in nor rejected them. They were just there, my own personal gang of five. There was no internal battleground or struggle. To be with myself at this time, happy and content with the existing moment, was all that I had and it was enough.

I smiled quietly at the first five stanzas, guiding me to let go. I was totally refreshed by the last two stanzas about living my days deeply in mindfulness and offering love and joy to loved ones to alleviate their suffering. I felt the universal nature of this wonderful benediction for both the living and dying.

The Five Remembrances focused my attention on impermanence; on growing old, getting sick, dying, losing loved ones, and realizing that my only possessions are the consequences of my actions. The final two stanzas of the meditation show the way; to live mindfully in each moment and offer joy to loved ones. As I practiced this meditation, I felt that each moment of life was absolutely precious. Somehow, I was communicating with all that I connected with. Before I slept that night, I felt my teachers and guides throughout lifetimes gathering together inside and around me, without boundaries. They stayed there while I slept and I was content and happy.

The next morning, to my surprise and joy, I woke up. Over the next six months, I slowly recovered my health. Friends in North America had, that December, booked airline tickets to take me out of India to recover in their home. I was touched by their love, but gently refused after thanking my friends for their loving concern. Whatever the outcome of this particular journey, it was to be in India. I'd written countless Christmas cards to friends and loved ones all over the world and signed them: *"Blessings and Love from Ian."* That is what I wanted to send out before my death. Then I lived, and was happy that the cards were sent.

The second circumstance that opened so many doors had to do with the shattering of my concepts about reality on an almost daily basis. I would have perceptions and judgments about a situation, person or event and would rapidly discover that they were without foundation. They were replaced by further perceptions and judgments, also without foundation. I felt a depth not previously known. This is something I call upon now when perceptions and judgments crowd into my consciousness. This willingness to let go of concepts or being with my body put me into a different space. I felt unseen hands guiding me through a year of initiations, mind training and transformations. I felt very privileged to receive the traditional wisdom of India.

Yet, I made it difficult for myself, with self-doubt, purification struggles and stringent endeavors to get it right. It was actually so much simpler than that, just being present with what is there. My happiness and delight came through being with humanity, the planet and universe, and serving all with joy. I did get caught, at times, in the process of struggle and purification. Then, for no apparent reason, the veils of illusion dropped away. A natural, overflowing delight in being and serving arose spontaneously.

I know I can never be as I was, nor do I wish to. I am simply grateful for all the gifts of transformation received. I also wonder about sharing these deeply personal experiences. I do not hold on to them, simply observing their effects on particular steps I took to tame my wild mind. The sharing is to illustrate that my approach to life comes through experience, crises, difficulties and joys that may have common ground with many readers.

If I can take steps along the spiritual path, surely anyone can. To the best of my ability, I endeavor to follow Gandhi's principles of *ahimsa* and the teachings on mindfulness. These are the guidelines and foundations for my peace and environmental activism. I eat a mostly vegetarian diet and live very simply as a planetary activist. Are there seeds of anger in my consciousness after all of this process? Of course they are. It is my job to ensure that I am not overwhelmed by their energy, but that I embrace the seeds of anger with the tools and practices I have received from my teachers. I observe how seeds of anger manifest in my thoughts, knowing they are capable of doing damage to me and to others. My practice has changed somewhat over the past three decades. It is not so much a focus on anger and violence but an observation of the tricks of ego.

My daily practice now is observing how my ego attaches itself to specific mental formations in order to take my consciousness into separation and illusion. That is the job of the ego. It cannot do anything else except attach itself to negative mental formations and distort and manipulate in order to separate me from my true nature. When I catch this happening in a train of thought, and I do not always catch it, I say:

Hello, my dear ego. Are you here again? Are you not tired of attaching to these old mental formations that you have used so often before? Why don't you come and have a rest in the consciousness of my heart?

The ego really has no answer to this. When I do catch it, I feel happy, really good, as the excesses of my wild mind are not translating into actions that can cause harm.

While in India, I also undertook two 28-day retreats, six months apart. They were the culmination of my training.

My cultural and religious background was not the same as my two cohorts, yet the experiences we shared were remarkably similar. I would observe my mental states, compare them with reports from my peers, and then discuss them with the *Swami* overseeing the training. Prior to the training retreats, I had months of preparation with attention to specific meditations, dietary regime and sexual abstinence. I learned how to chant the *Gayatri Mantra* and co-ordinate it with the four components of breath: inhalation, holding the air inside, exhalation, holding the emptiness. There was a mathematical precision in tone, pitch and resonance of the mantra, as it was exactly co-ordinated with the different components of breath and hand movements over the body. It was complex and overwhelming and I frequently wondered if I would ever get it right. I benefited enormously from the encouragement of my cohorts who were determined that I would not be left behind. I also had skilled and patient teachers who made the effort to transmit this oral tradition, thousands of years old, to a westerner not used to this form of education.

The second training period in a different part of India, Karnataka as opposed to Andra Pradesh, was with a new cohort made up of experienced meditation teachers and exceptional gurus. With this powerful group of beings, the sunset ceremony was conducted by running water to deepen the silence, stillness and penetration of the mantra. The chanting of the *Gayatri* took place with all of us standing up to our waists in the water. When it came to the point of suspending thought and allowing the *Gayatri* to arise spontaneously, to my total astonishment it did just that. At the same time, I could feel and identify the particles of mud between my toes, see minute electrons in the air and look down on my wisdom buddies from a great height. I felt encompassed by the evening sky and at the same time I encompassed the sunset, the evening sky and everything beyond it. This experience was repeated with varying intensity during every sunset rendition of the *Gayatri Mantra*. I never felt it necessary to communicate this to the *Swami* or to members of my second cohort. I went into total silence and do not recall talking to anyone, as everyone very carefully left me there.

In my diaries, I recorded my experiences in poetry and art, a totally inadequate exposition for something that cannot be fully expressed in either. I persisted, despite this inadequacy, to convey some semblance of the experience through words. Before I took my leave from the ashram, the *Swami* asked to speak to me. He described my experiences in complete, precise detail and arranged a parting ceremony, an initiation to acknowledge the grace of a guru now recognized with the name bestowed: *Prem Chaytania*. My cohort was delighted by this. Training with *Gayatri* had major life changing effects, not the least being that I became a better and more skilfull teacher, both in meditation and to university students.

What I can say from personal experience, is, once my wild mind was reined in, clarity and compassion were suddenly there in greater compass, providing a different, new basis for how to be with the planet and others.

This partial account of my journey in India is to demonstrate that my activism for peace, planetary care and social justice now come from a different place as a result of my internal work. Steadiness, clarity and compassion are within me, rather than ego posturing from the lunatic fringe. Though there was a "rush" from the latter, I prefer the still-point, uncoloured by the excess of ego and desire for recognition. Such a still-point permits me to be free in my own sovereignty, no matter what I am doing. It also propels me to serve the planet and humanity by creating bridges and pathways of harmony.

The rest of my life is still a work in progress.

13: **Prologue: Chronicles of Awakening**

Book One: Redemption

Book Two: Trailing Sky Six Feathers

Book Three: New Planet, New World

New Planet, New World is the final book of a trilogy that was completed in 2016. The first book – *Redemption* – was a lost manuscript first written in 1975. It was soon forgotten, as back then I didn't know how to get published. This heartfelt novel was rediscovered by accident in 2011. I found it in an old filing cabinet where it was gathering dust. I went through it and could scarcely believe it was such a good read. I requested my wife and friends read it with critical eyes, just in case I was dreaming.

One friend cried all the way through, the other mused about the film to be made. Modern technology enabled the yellowing typed manuscript to be transformed into a computer-ready document. The narrative still stands pristine as when first written 40 years ago. The story is an allegory for the life difficulties I experienced at that time. The surprise for me was how could I have written such a book about awakening, while in a desperate state of mind? I was in a bad place with a failing marriage in the Hebrides, Scotland, while trying to create a career at Carleton University in Canada. I was not doing a good job with either.

Redemption is set in The Hebrides, islands off the northwest coast of Scotland, with startling cycles of maturing and downfall for the epic character, Callum Mor.

He was a gifted child, master mariner and derelict drunk, who eventually gains wisdom from a hard life's journey. He enters the dark zone of alcoholism and withdraws from society. With only his animals keeping him this side of sanity, he survives in a bleak solitude.

Laced with grim humor, the novel has nature's harsh and beautiful rhapsody as the background for tragic human failings; violence, power, murder, rape and madness. The failings are ultimately crowned by the triumph of the human spirit.

A family with a young girl seeks refuge from a storm at his house and slowly Callum Mor steps away from self-destruction to an astonishing awareness that triumphs over his tragedies. He saves the girl's life in a blizzard and the glimmer of awakening dawns in him, setting the stage for the final drama that illustrates the resilience of the human spirit.

Redemption is a deeply moving tale of desolation, love, loss, transformation and hope. It reads like an extended prose poem, reflecting the primal forces of nature and of humanity. Its starkly gorgeous and remote island setting creates and reinforces the central themes of struggle, family, community and wonder at the beauty of the world. The rich cast of characters offers interludes that brim with interpersonal drama.

The story centers on and is always connected to Callum Mor, but he is surrounded and influenced by a fantastic cast of family and fellow islanders. They provide a deep well of material, as their conflicts and intrigues move the plot forward and offer a vast array of powerfully emotional moments. The story arcs of other characters in the novel offer intriguing counterpoints to one another and to Callum Mor. Their hopes, desires and difficulties intermingle in a tapestry of human existence.

The narrative tone is generally quiet and introspective, but it is frequently punctuated by storms, both literal and metaphorical. Loaded with the symbolism often found in parables, *Redemption* alludes to more than what is openly stated. Every scene provides a striking visual clarity that slips into the realm of timeless storytelling. All of this enables deeper, more subtle messages of compassion and faith to carefully unfold. From the rhapsody of an idyllic childhood, through traumatic tragedies, to the derelict zone of alcoholism and then a state of awakening, I depict the stations of a personal Calvary that ultimately leads to *Redemption.*

Dr. Tom Hagen, his wife Sian and his daughter Catriona comprise the family taking refuge at Callum Mor's house. They are writ large in the final book. I place them in the near future of 2080. Dr. Hagen becomes the chef-de-mission of the International Space Agency's mission to settle on a planet in a nearby galaxy. Tom, Sian and Catriona move from minor characters in Book One to the main ones in Book Three, *New Planet, New World.*

Book Two of the trilogy, *Trailing Sky Six Feathers,* is a hero's journey. Almost Indiana Jones meets the Buddha with a dash of Celestine prophecy. Shamanic healing of childhood sexual abuse, guru training and a near death experience in an Indian ashram are all part of stumbling through the first stages of life, then standing strong in my own sovereignty in the latter part.

Past life memories collide head on with the present, all thanks to the persistence of Trailing Sky, the Muse who refused to give up. Karma is reversed; the internal battles are over as the author begins to live life through meditation for Gaia. The relentless shadowing by this engaging Muse brings understanding not only to me, but to anyone involved in overcoming the darkness of their past.

With a voice steeped in authentic experience, I navigate past and present lives over four centuries; from brutal raids on Indigenous settlements in 18th century Arizona, rough sea voyages off the Scottish Hebrides in the 20th Century, to a decisive life moment of surrender to the Muse in the 21st century.

These tales weave seamlessly and can create inspiration for a wide range of fellow spiritual seekers. The genre is legend mixed with autobiography.

Trailing Sky initiates a dream vision in 2008 that caps my process of remembering a mosaic of experiences going back four centuries.

Over a period of 30 years (1980 - 2010), four extraordinary mentors enhanced this process of remembering for me, while Trailing Sky waits patiently in the distant past.

I learned to reconfigure my understanding of time, place, consciousness and Carl Jung's psychology.

When I talk to people about Book Two, the first question is usually, "Why did you write this book?"

I explain: "Global Citizens are staring into the abyss, yet instead of being eaten up by it all, I say to you 'awaken spiritually.' That changes everything. We have made our world an unpredictable beast because we fail to work within it.

We have to take back control of ourselves and this is a spiritual matter to take control of our impact on the ecology of Earth.

Turning on the switch of awakening is a good idea right now. We need to touch the sacred in ordinary experiences of life to find the courage, skill and determination to transform.

I wrote *Trailing Sky Six Feathers* to shed light on issues that will affect our world for generations to come. The example of my own challenging journey and personal transformation illuminates one path to inspire others to choose their way to expand their consciousness and chart the course for the future."

I conclude noting: "Humanity does not need to continue with maladaptive options and patterns. We can and must transform. The key to changing this is awakening a spiritual relationship with self and the Earth."

Our industrial civilization is a system devouring itself, dislocating the organic structures of Earth to the point that all species, not just our own, are at risk.

It has taken us to a dangerous precipice. From there we stare at climate change and ecosystem and financial collapse. These lead to humanitarian concerns such as ISIL, resource wars, cyberbullying, terrorism and anarchy.

The two main characters that open Book 2 in 18th century Arizona are Trailing Sky Six Feathers and Eagle Speaker.

When the reader meets Trailing Sky Six Feathers, my Muse from the past, they encounter a powerful, relentless woman who transforms my life in the 21st century, not in historical fiction. She has been described as one of the most powerful women in modern Canadian literature.

Eagle Speaker is her husband and also my transformation vehicle. He dies cradled in her arms in a medicine wheel in the year 1777.

As he takes his last breath, Trailing Sky whispers to him, "I will find you. I will find you." She assures her daughter, Rising Moon, that she too will find him.

Rising Moon has a minor role in Book Two, yet by transferring her to the new planet in the final book, *New Planet, New World,* I bring the 18th century to connect with the 21st century. Time, culture, space and consciousness are fused across centuries to create the final book of the trilogy.

New Planet, New World provides a counterpoint to the demise of modern civilization. I chart a beginning anew for humanity, a communal hero's journey to reconstruct society based on ecology, caring and sharing.

Powerful elites ignore their complicity in the destruction of life on Planet Earth.

This adventure is not without risk or cost. The clash of centuries opens Chapter One with a dangerous meeting on a distant planet later this century. The protagonists are from different centuries and cultures.

From the 18th century, Rising Moon is hurled by shamanic means to Planet Horizon in a nearby galaxy.

From the 21st century, Catriona gets there in an escape craft from a failing spaceship. Catriona is taken prisoner but fights back, screaming, "I am not your enemy." Instead of killing one another the two young women choose to be blood sisters and embrace survival, accepting nature as their matriarch.

This fragile thread is challenged by the brutal abduction and rape of a main character, Sian the Celtic seer. Her inner strength, of being more than a violated body, inspires the community of pioneers who had escaped safely from the damaged spaceship.

They create a communal structure of living and carve out a home and presence on the new planet.

Four Hopi Sacred Keepers offer their lives in a ceremony to enable renewal on a distant planet that none of them will experience.

Mysticism combines with technology to enable a transfer particle to seed the new planet and establish settlements.

The expansion of communities is interrupted by a jihadist attempt to take over.

A terrorist cell on Earth hijacks a spaceship and imperils the lives of the pioneers, who respond with tactical violence to kill them. The stark violence of survival prepares a backdrop for three distinct love stories to emerge.

Ethical settlements grow as a mirror for Tolstoy's vision of "people of the twenty fifth century" – ahead of their time.

The dark episodes and lighter passages move the story along with action, fear, resolution, death, execution, rape, bravery and exile in a futuristic opportunity for humanity.

This action packed book of intertwining plotlines arc into the epiphany of the final chapter, which muses about human survival anywhere. This end game is a philosophy for the future.

The inclusiveness of science combines with Tolstoy's vision, Pope Francis' Climate Change Encyclical and not repeating the mistakes of the carbon cabal.

The underlying message is from Tolstoy, the 'Conscience of Humanity.' He described humanity's bottom line as the cultivation of love, the mainspring for authentic and responsible living.

I do not present this as idealism, rather as basic wisdom. That is why I wrote this futuristic novel, the final book of "Chronicles of Awakening."

> *"If I can take steps along the spiritual path, surely anyone can. To the best of my ability, I endeavor to follow Gandhi's principles of ahimsa and the teachings on mindfulness. These are the guidelines and foundations for my peace and environmental activism."*
>
> **- Dr. Ian Prattis**

PART FOUR:
SPIRITUAL SUPPORT

"*In its beauty, grandeur and compelling intimacy, the mystic ultimate dimension of the Buddha dispelled my scepticism. As I read different chapters of the Lotus Sutra, I was transported to the worlds and dimensions described. I would read a little, and then put the book down as I felt myself going deeply into meditation. I was profoundly moved by the content and experiences through the series of translations into Chinese and then into English.*"

- Dr. Ian Prattis

14: Dawson's Desert Legacy

Dawson was a wisdom holder of many traditions – Ojibwa, Hopi, Lakota and the Native American Church. He did have a second name, but preferred Dawson. He was a legendary figure in Central Arizona and left a lasting impression on everyone he met.

I have encountered many people at conferences and talks all over North America and when it emerges that I have spent a considerable amount of time in Central Arizona desert country, I am always asked if I know a man named Dawson. He had met all kinds of people in his capacity as a guide and teacher. Yet his attention and presence never wavered in its intensity as he welcomed all into his orbit of wisdom and patience.

I first met Dawson in 1987 on a day long ethno-botany field trip he offered in the Sonora desert region of Central Arizona. I was the only person to turn up, yet this did not deter him. He generously extended his knowledge of plants and hidden sources of water in the scrubland of the Sonora desert. His field trip skirted ancient medicine wheels created centuries ago. He talked about plant cycles within the teachings of the medicine wheel both for ceremony and healing. His mentorship has always meant a great deal to me, especially his instruction of how to build a medicine wheel.

Dawson was a slender yet muscular man in his sixties, though he seemed much older. His manner was slow and deliberate, gentle but firm though his light blue eyes carried a steely glint. He loved movies and would always sit in the cinema until the end of the credits, always the last person to leave. Eyes closed, he made a point of downloading the full feeling of the film. It was the same with people, animals and the desert.

He brought a sense of gentle intensity and intimacy to every relationship. The initial connection from that first field trip and movie experience warmed into a friendship. One evening in Sedona, two years after our initial meeting, I received a call from him. He asked if I would pick him up two hours before dawn the next morning.

"Wear hiking boots," he said.

I drove in the early morning dark to Cornville and found him waiting outside his house. I followed his directions to take various forestry roads leading to a reserve on the northern fringe of the Sonora desert. After parking we hiked for approximately thirty minutes into the desert through a scrubland trail.

It was still dark when he gestured that we should stop. We shared a flask of coffee and the intense silence of the desert, interrupted only by the scurry of small wildlife. In the dark of morning just before dawn Dawson gestured for me to look in the direction of three large cacti directly in front of us. The sun rose and I could vaguely make out the flowers opening. Then Dawson pointed them out. They were absolutely stunning in their unreal beauty, ranging from yellow to dark violet. We sat there for over an hour, appreciating their beauty, as the morning sun rose.

"You had to see this before you travelled home to Canada," were his only spoken words. The morning heat was suddenly broken by a sudden hail storm. We put our packs over our heads and ran quickly to the shelter of the nearest rocky outcrop. The storm lasted only ten minutes although the stones were not small, making quite an impact on any unprotected area of the body. Dawson looked at me strangely.

"That sure is some kind of acknowledgement from the past, and it isn't for me. What have you been up to Mister Ian?" Dawson asked.

I just shrugged, as I had no intimations of cause. We walked in silence to where I had parked the car. The hailstones were not to be found beyond a hundred yard perimeter of where we had been sitting.

"Beats the hell out of me, though I reckon you will have some building to do back in Canada," said Dawson cryptically, as he peered at me out of the corner of his eye. These were the last words I heard him speak. As was his custom we drove in silence. He got out of the car by his property, waved once and was gone.

On a later journey in 1992 to that region of Arizona, when enquiring about him, I discovered to my dismay that he had been killed in a car accident outside Phoenix. I was deeply saddened by this loss, thinking about all that he had so patiently taught me.

I drove to where I had last walked with him, to pay my respects to this extraordinary spiritual teacher, remembering the way almost without thinking. It was not the time for the cacti to flower but I treasured once again the gift he had shown me. I wondered who he had passed on his vast knowledge to, then realized suddenly that he had passed on a great deal to me about medicine wheel lore and construction. Dawson was a spiritual guide and had taken me through many shamanic journeys. The hailstone storm was no longer a mystery to me, rather an early prompt. What I had received from him was put into place in the hermitage where I lived, in the Gatineau Forest in Quebec.

Over a period of five months in the spring and summer of 1994 I experienced very intensive shamanic journeys with an Algonquin shaman that I prepared for through fasting, meditation and sexual abstinence. On five separate journeys I met and dialogued with ancient shamans from the East, the South, the West, the North and finally to the ancient shaman of the Center.

I figured at first that this was an experience with five facets of the same archetypal material from my deep unconscious, though there were major surprises I had not anticipated. Each shaman created distinctive unconscious energy within me, interconnected to the other four. In each journey I was always met by the same beautiful female figure, who then led me to the ancient shaman. Dawson had repeatedly told me that the feminine source would eventually emerge as a Muse for me, and there she was.

At my hermitage in the middle of Gatineau Park Forest in Quebec, I had a small circle of large stones in my front yard with beautiful ferns growing at the center.

I had an overwhelming compulsion that summer of 1994 to build a medicine wheel with this circle of stones as the interior circle.

I had been taught by Dawson the appropriate mind-state and procedure of respect to construct a medicine wheel. Dawson had instructed me intensely in Arizona about the central circle of the medicine wheel. It could only be truly experienced when connection to the sacred mystery was intact.

The four cardinal directions, East, West, South and North, were the organizing axis for this ultimate fusion, represented by the ferns over which I took such care. It had sunk into my intellect but now reached my heart.

I constructed the medicine wheel with the assistance of two friends who shared my respect and training.

We carried out the appropriate ritual, and worked with reverence on a very hot and humid summer's day. The silence that settled on all three of us spoke of something happening inside and around us while creating this architecture of incredible grace, power and beauty.

The stones for the medicine wheel came from my garden and the surrounding forest, the hard granite of the Canadian Shield, part of the very ground where the medicine wheel was being built.

After filling the four quadrants of the medicine wheel with fresh garden soil, we contemplated what had been created.

I realized its connection to my five shamanic journeys over the previous year. The cardinal points of the wheel and its center were a reflection of the five ancient shamans I had journeyed to meet and the ferns at the centre were an appropriate symbol for the feminine muse that delivered me.

The medicine wheel was a symbolic map of my internal experience.

I was re-inventing the wheel from my journeys to meet the five Ancient Shamans, yet also ensured that the beautiful ferns remained intact at the centre of the medicine wheel.

Now I started to smile at how this medicine lore and knowledge had gradually seeped into my consciousness from Dawson. His overarching influence had prepared me for the journeys to the five shamans.

I could feel his intense blue eyes watching me at this moment and perhaps he permitted himself a smile too.

It was his instructions I followed for my medicine wheel. He had known that I would eventually understand the wheel and the space at the center as the locale where I would seek counsel from the internal feminine - the beautiful ferns.

"It was into silence, deep silence and stillness amidst the world I lived in. This is where I found the causes and conditions that would provide tidal waves of energy to my cells and consciousness. I could truly look deeply into my suffering, into the dark areas that held my mind hostage with formations of an unwholesome nature. Over the past decades, I have built more and more silence into my everyday life."

- Dr. Ian Prattis

15: Sound of Silence

Paul Simon wrote *The Sound of Silence* in 1963 and with Art Garfunkel, recorded this song with Columbia Records a year later. It failed and led to the duo breaking up.

Later on, the song's producer, Tom Wilson, did a remix of the original track, overdubbing electric rock instrumentation.

The song became a Number One hit all over the world and brought the very surprised Simon and Garfunkel back together.

They were university students and part of the counterculture movement, yet Simon had no intent other than writing a good song in his bathroom while he played his guitar with lights off and the water running. He was only twenty-one years old.

Garfunkel provided a focus on the inability of people to communicate but it seems as though the lyrics wrote themselves.

It took the American heavy metal band "Disturbed" and their lead singer David Draiman, in 2015, to add a sharper edge. Their rendition was not just great music; it was a cry of pain for our entire civilization.

The lyrics of Simon and Garfunkel's *Sound of Silence* are insightful about society and the planet. The imagery shone light on humanity's inability to communicate with any harmony, the "neon god" no less.

"People talking without speaking

People hearing without listening

People writing songs that voices never share."

Note the enigmatic ending:

"The words of the prophets

Are written on the subway walls

And tenement halls

And whispered in the sounds of silence."

Does this sound all too familiar? Whether Simon and Garfunkel recognized it or not, the song is highly provocative in the awakening process. The lyrics carry a steady context about the necessary expansion of silence. They provide a message that until there is silence, there is no place for the wisdom of the prophets to penetrate human consciousness. The latest version of this masterpiece, by Disturbed, places it in our current societal and planetary collapse. I extrapolate on the deeper significance of this overlooked aspect of Simon and Garfunkel's song and draw on two individuals from the realm of prophets. I refer to the Buddha and to Ramana Maharsi, and explain my limited experience for good measure.

I take a more intense tangent on silence with the Buddha and Yasoja. Ten days before the rainy season retreat, Yosaja and his five hundred monks journeyed to where the Buddha held his three month retreat. They arrived in a boisterous way to greet the monks there with loud greetings and lots of talking. The Buddha heard this uproar and asked his faithful attendant Ananda, "What is that noise?" Ananda replied that the venerable Yasoja and his followers had arrived and were greeting the resident monks. The Buddha asked for them to come to him, so he could send them away and dismiss them for their noise.

The five hundred monks and their leader bowed to the Buddha and left the rainy season retreat in Jetta Park. They walked for many days to the east side of Koshala and arrived at the Vaggamuda River. Once there, they built small huts to begin their own rainy season retreat. Yasoja addressed his followers and told them that the Buddha sent them away out of compassion, so that they would practice deeply. All the monks saw this as true and practiced very seriously to show the Buddha their worth. The majority of them realized levels of enlightenment during their three month retreat. The Buddha's rainy season had also finished and he remarked to Ananda that he could discern the energy of goodness and light emanating from the east. He realized that Yasoja and his five hundred monks had achieved something very deep and sent them an invitation to join him.

They arrived quietly in the evening after many days of silent walking to find the Buddha sitting in silence, in a state of concentration called imperturbability – free and solid. When they saw this, they decided as one body to sit like that with the Buddha and entered the same state. Ananda approached the Buddha during the three watches of the night and asked him to address the monks. The Buddha remained silent. After the third reminder he said, "Ananda, you did not know what was going on... I was sitting in a state of imperturbability and all the monks did the same and were not disturbed by anything at all." In this deep, unshakable silence, the communication between the Buddha and Yasoja's five hundred monks was perfect. A deep transmission of insight, freedom and joy went to them. No ceremony was required as the monks experienced a natural awakening – all from imperturbable silence.

During my yogi years in India, I had the privilege of training in Sri Ramana Maharsi's tradition through Siddha Samadhi Yoga.

I had been recognized as a guru and taught meditation in Mumbai and Bangalore. I made a point of staying at Ramana Maharsi's ashram near the holy mountain of Arunachala in South India where he stayed until his death in 1950. I followed his footsteps up the mountain and meditated in the cave where he first took shelter. Bit by bit, I entered into his zone of silence, though he was long gone in body. Yet it was of the same nature of imperturbable silence as described in the Buddha's welcome of Yasoja. Sri Ramana emanated the same force of freedom, which stilled the minds attuned to it. He offered a transmission of the state he was perpetually immersed in that could be directly experienced by those sitting with him.

This was his preferred method of teaching, though he would verbally address the issues and questions brought to him by students and followers from all over the world. His verbal teachings were there for those unable to understand his silence. He provided guidelines to practice a vigorous method of self-examination: 'Who Am I,' 'Whence Am I' – to help them step into the silence of their true nature and experience that consciousness alone exists, giving their thought tortured mind a rest. His simplicity, humility and sense of equality were legendary.

He always shone like a beacon as he had realized that his real nature was unrelated to his mind, body and personality. He was accessible to everyone, shared in communal work at the ashram and rose at 3:00 am every day to prepare food for visitors – always eating last, after everyone had been fed. He lived, slept and held audience in the small hall of the ashram. I used to sit and meditate there during my stay and could feel and imagine how he would address the questions of the constant flow of visitors and at the same time radiate his silent presence.

His spoken teachings all arose from deep in his heart - from direct experience that consciousness was the only existing reality. It was through silence, that his disciples would know the same.

It was the depth of his heart that moved people, which demanded only the exit of ego and the entrance of trust in the arising consciousness and patience for the flow. That threshold is what moves the other into the space of origins, making it feel authentic.

We are surrounded by a modern, noisy, ungrounded world that opens so many avenues for disaster, yet Sri Ramana Maharsi ably demonstrated that there are conditions to change the possible disaster into transformation. That is how I endeavour to think, write and speak. This brief reference to Buddha, Yasoja and Ramana Maharsi describes universal consciousness.

When Thich Nhat Hanh ordained me as a dharma teacher, he transmitted the Lamp of Wisdom in a ceremony at Plum Village in France. This was in 2003. I was required to present a dharma talk to the monastics present on this occasion. I talked about waves and water in order to come around to the significance of silence.

This is what I said:

My teacher, Thich Nhat Hanh, uses a wonderful analogy of waves and water to understand how the historical and ultimate dimensions of reality are interwoven. Waves rise, fall and die when they wash up on a seashore or riverbank. This is the analogy for the historical dimension, our daily existential cycle of life full of crises and ups and downs. But no matter what attributes apply to waves, there is always a constant. While a wave is high or low, born or dying, coming or going, it is always water. The constant of water refers to the ultimate dimension.

The idea is that if we understand the waves of life, then we can touch the water of life - the ultimate dimension. It is a transcendent reality, outside of time and space and distinct from the constraints of our daily existence. We can call this Nirvana or the Kingdom of God.

I have heard Thich Nhat Hanh explain the waves and water analogy many times. The metaphorical qualities made sense to me. But deep looking into my waves did not lead me to touch the water of the ultimate dimension.

My 'waves' did not bring me to the 'water,' as I certainly expected them to do after listening to my teacher. I wondered for a long time about this disjunction between my acceptance of this notion and my lack of personal experience. There were three logical options to investigate.

1. The first option was that Thich Nhat Hanh was incorrect.

2. The second option was that Thich Nhat Hanh was neither correct nor incorrect. He was simply very generous in choosing not to chart the difficulties of transition from waves to water.

3. The third option was that Thich Nhat Hanh was correct and that something crucial was missing from my practice.

I eliminated the first option as I have great trust and faith in Thich Nhat Hanh. There may be something to the second option as I know how generous he is, that he may choose to encourage rather than chart the difficulties on the path. Yet, I realized very early on that the real investigation was the third option - to find just what was missing from my practice of mindfulness.

I was aware that my waves were too small to carry me through to the ultimate dimension - too small in terms of insufficient concentration, insight and mindfulness – the three energies of transformation. What I needed was a tidal wave to make my waves full so that this energy could provide the "voltage" to make a transition from waves through to water. My investigation turned inward, to the causes and conditions that would make my waves into tidal waves full of concentration, mindfulness and insight. As I pondered this deeply, I stumbled across where I had to go.

It was into silence, deep silence and stillness amidst the world I lived in. This is where I found the causes and conditions that would provide tidal waves of energy to my cells and consciousness. I could truly look deeply into my suffering, into the dark areas that held my mind hostage with formations of an unwholesome nature.

Over the past decades, I have built more and more silence into my everyday life. On a daily basis I stop, look deeply and dialogue with the seeds in my consciousness – a practice received from my Native American medicine teachers. My consciousness was guided by these seeds of awareness, to transform difficulties and impediments in my life, enabling me to move on.

My home and sangha life, supported by the entire Pine Gate Community, enables me to retreat into silence on a regular basis. In this way – through silence and deep looking – my waves became bigger, more infused with concentration, insight and mindfulness. Deep silence and dialogue with the internal feminine provided the conditions for my waves to become tsunami. As I continued to rest in the silence and look deeply into my shadows, there emerged the distinct experience of touching the water. Thich Nhat Hanh was correct. I had to discover for myself the significance of silence and skillful deep looking.

The fruits of this practice of silence and non-action were many and particularly manifest in my study of the Lotus Sutra.

I applied myself to study the Lotus Sutra, particularly Burton Watson's 1993 translation from the Chinese version by scholar-monk Kumarajiva in 406 CE. Prior to this intensive study, I was much more comfortable with accepting the Buddha in historical form.

The story of the Buddha's life, awakening and ministry was enough for me and I had not paid too much attention to the Buddha in the ultimate dimension. That changed radically through reading the Lotus Sutra and from my practice of silence. For in the Lotus Sutra, the Buddha in the ultimate dimension is revealed in no uncertain terms.

In its beauty, grandeur and compelling intimacy, the mystic ultimate dimension of the Buddha dispelled my scepticism. As I read different chapters of the Lotus Sutra, I was transported to the worlds and dimensions described. I would read a little, and then put the book down as I felt myself going deeply into meditation. I was profoundly moved by the content and experiences through the series of translations into Chinese and then into English. I would remain in a trance-like state for hours.

My direct experience of the energy of this Mahayana masterpiece brought forth many insights. The most pertinent one was that I would not be able to experience the Lotus Sutra in this way if my waves were still too small, lacking in insight, concentration and mindfulness.

Without the silence and what it enabled, I am sure I would have had a superficial reading of the Lotus Sutra that would not have allowed me to touch its depth and magnificence.

The Lotus Sutra is full of the activities of bodhisattvas, sages and holy beings, and of how we may understand their roles. The bodhisattvas are described as being immersed in the ultimate dimension, and from there they return to the historical dimension to transform suffering. As 'water,' bodhisattvas live the life of a 'wave.' Choosing to do so encourages us to come face to face with suffering, to step away from fear and take our own path into freedom.

This is the task of the true revolutionary of the twenty-first century. Not to pick up a gun and shout hatred, but to penetrate 'water' from the 'waves' of life.

There are so many bodhisattvas, from all spiritual traditions, who are choosing to do this. In a way, this ushers in the end of religion – of being attached to the identity gained from one's religion. The task before us in the twenty-first century is to step out as spiritual warriors.

We should not be caught by our religious identities but connect and walk hand in hand with friends from other spiritual traditions who are doing the same. I am expanding the term bodhisattva so that it embraces far more than Buddhism.

I came through this process with waves that are not so small anymore and a full heart to share with everyone. I also experienced a cycle of internal interconnectedness.

Empowered by my study of the Lotus Sutra, I institute yet more silence into my life even when I am talking to someone or even offering a dharma talk.

I became available in a manner I was not before. My waves carry more voltage and are filling up rather than being half full. My activism for peace and the environment rests on a foundation of silence and the initial necessity of imperturbable stillness. The true art of doing nothing.

The Sound of Silence
(Paul Simon)

Hello darkness, my old friend
I've come to talk with you again
Because a vision softly creeping
Left its seeds while I was sleeping
And the vision that was planted in my brain
Still remains
Within the sound of silence
In restless dreams I walked alone
Narrow streets of cobblestone
'Neath the halo of a streetlamp
I turned my collar to the cold and damp
When my eyes were stabbed by the flash of a neon light
That split the night
And touched the sound of silence
"Fools" said I, "You do not know
Silence like a cancer grows
Hear my words that I might teach you
Take my arms that I might reach you"
But my words like silent raindrops fell
And echoed in the wells of silence
And the people bowed and prayed
To the neon god they made
And the sign flashed out its warning
In the words that it was forming
And the sign said "The words of the prophets
Are written on subway walls
And tenement halls
And whispered in the sounds of silence

16: **A Manifesto for the Future**

As a Zen teacher I make a commitment not to cause harm. I am guided by spiritual ethics yet am aware that the current disastrous state of the planet will not bring forth strategic plans of how to fix things.

I could go on and on about the terrible things taking place in society and to the planet – and will divert to that in a moment. Yet the bottom line for me is to remember and refine a system of ethical conduct. I go deeper into meditation and fix myself to be steady and insightful. I register with Mindfulness Trainings, as it brings out all that I would like to see in people around the planet.

The bottom line for me is that awakening and mindfulness are active. Activism, on its own, does not have the inner resources to maintain effective social and planetary transformation. I know from personal experience that re-training the wild mind is a necessary ingredient to precede activism. Becoming environmental or political is only one part and cannot be fully effective until the internal side is in place.

We have no alternative but to concentrate on sustainable living, rather than exploiting the spoils of perpetual economic growth. Profit cannot be the sole reason for commerce. There must be responsibility tied into the equation. At present, we are totally out of sync with the earth's resources. The fragile threads of ecosystems around the globe are severely compromised. We are in the position of either going down the collective sewer or changing our values in the direction of awakening.

Jane Goodall issued a dire warning in 2016 that 'life is hanging by a thread,' as all living things will be negatively impacted by rapid climate change.

In particular, she advocates the necessity of creating programs that stop tropical deforestation by making rural communities custodians of the forests.

This is difficult when President Trump has begun to dismantle environmental regulations, setting in motion irreversible consequences around the world. The United States is ignoring climate change, obstructing clean energy and many forms of conservation. Noam Chomsky in 2016 refers to Trump's priorities as "…racing as rapidly as possible to the destruction of organized human life."

Stephen Hawking's thoughtful piece in the Guardian (December 1, 2016) places a focus on elite behavior creating further inequality as he examines Brexit and the Trump presidency. His question is how will the elites change? "We are living in a world of widening, not diminishing, financial inequality and people see only a slim chance at earning a living at all." Hawking acknowledges this dangerous moment in humanity's evolution.

Earth is like a giant living cell, all parts are linked symbiotically. Biologist Thomas Lewis created this metaphor with humanity just as one part of a vast system. This is not something that powerful and corporate people have paid much attention to. The reality is that the life support systems of the planet are severely threatened by climate change, aided by accelerating global consumerism. Our ignorance and neglect are destroying Earth, because we do not know how to respect ourselves, others, and the planet. Unless we radically change, there is no possibility of balance, environmentally or socially. This became clear in my filmed distance course "Ecology and Culture" at Carleton University in Ottawa, Canada. I wanted to connect the many levels of violence and fear we engage with to the environment, and to the everyday use of harmful speech and mindless consumption.

With ethical guidelines rooted in spiritual practice, we do not generate the energy that enables terror and violence to grow. However, everyday situations become swamped by the overall climate of fear, hatred and vengeance. We need to learn to behave differently.

These issues were examined with great clarity by the awakened mind of the Buddha, 2600 years ago. His teachings are timeless, as relevant to the modern world as when first spoken.

The Buddha taught the Five Mindfulness Trainings as a design for living. Thich Nhat Hanh reworked them to relate to modern realities. They are non-sectarian and all spiritual traditions have their equivalent. The first training is to protect life, to decrease violence in oneself, family and society. The second training is to practice social justice, generosity and not exploit other beings. The third is responsible sexual behavior for all people, to protect couples, families and children. The fourth is the practice of deep listening and loving speech to restore communication and reconciliation. The fifth is about mindful consumption, which helps us not to bring toxins and poisons into our body or mind.

I asked the students in my Ecology and Culture class if anyone would care to read them out to their classmates during my lecture on environmental ethics. There were many volunteers. I did wonder if this borrowing from Buddhism would go over well with students and the viewing audience. Much to my surprise, students and the public viewers wrote in to tell me that this was a wake-up call, the first time they had been presented with specific environmental ethics. Let me be clear, the trainings are not there to judge others. They are an internal guide so that, as individuals, we wake up to love and compassion and take heed of the directions the mindfulness trainings take us in. The trainings are not a coercive design for conformity.

They simply assist us to be more aware of what is going on, around and within us. They enable us to distinguish what is good for ourselves, our minds and the world and what is not. It is not necessary to complete the practice perfectly, as that is not possible.

It is, however, possible to move in the direction of responsible and ethical living and make a difference to our society and environment. Do we bring to violence, indifference and terror a renewed application of the same or do we step back and consider these teachings?

There is a solution to our present situation. Our leaders have often become trapped by corporate and electoral agendas, following a similar script, seeking justification and in some cases, avocation for the use of violence. Large scale change is difficult to find within this system, but the Buddha offers a path. The implications of his Five Mindfulness Trainings apply to the dangerous times we live in. Our world needs guidelines like these.

A flip side to global violence is the growing concern over the absence of love, decency and compassion in daily public life. This preoccupies and worries many citizens and scholars. If there was ever a time to learn anew from these teachings, it is now. When we touch base with the Five Mindfulness Trainings, we are being reminded to wake up. Neglect, terror and fear are states of mind. Therefore, we need tools that reconnect us to a mind-state driven by love, decency and positivity.

The Five Mindfulness Trainings are presented as an antidote to the contemporary crises. The ethics of the Five Mindfulness Trainings provide a necessary balance to find our true nature, while caring for all we connect with. In addition to addressing social and environmental crises, the building of inner spiritual strength through meditation and mindfulness is crucial.

However, I must point out that it is critical in the 21st century that necessary re-education also find a place in the Five Mindfulness Trainings. They are indeed a guidance system to encourage us to no longer participate in a non-sustainable economic system driven by greed and distraction. This global ethic is our protector as it helps us to stop, look deeply and throw away our harmful patterns of behavior. Crises such as Climate Change prompt us to refresh and refine the trainings but as we will see there were some awkward disconnects in their creation. This begs the question of how to relate to the trainings without a disconnect to their intentions?

The Buddha was clear about impermanence and new challenges. He created the Five Mindfulness Trainings for the lay community and told Ananda, his faithful attendant, that the minor precepts should be revised according to the culture and the time. But Ananda and the Buddhist elders were confused about which precepts were the minor ones and misunderstood what the Buddha was talking about. And so nothing changed for 2,600 years. There was no preparation or anticipation for modern realities, as monastic precepts haven't changed much and were not equipped to handle issues ranging from Internet, terrorism, a world full of refugees, to Climate Change.

The seeds of disconnect are not just with the trainings but with dharma in general. The disconnect reveals itself in terminology. Minor precepts refer to the Five Mindfulness Trainings for lay people while major precepts define monastic ethics. This language creates a divide between lay and monastic with the latter considered as superior, which is certainly not the case. In the modern era it is the lay dharma teachers who are in society, working in the trenches of everyday life, creating transformation in alliance with many other groups of lay people.

Whereas, the monastic community is secluded, often cut off from everyday reality and is not in a position to create transformation in the wider society.

This disconnect is a marker of modern Buddhism in the west and was noted by David Loy in his excellent article in Buddhadharma (Winter 2015.) Loy addresses the current ecological crisis and questions the deep rooted ambivalence within Buddhism towards it. He asks "Does the ecological crisis have nothing to do with Buddhism?" I add a further enquiry, "Where are the Buddhist politicians, CEO's, entrepreneurs in political, ecological and economic spheres?" There is a wide disconnect in Western Buddhism between playing the capitalist game, yet only being concerned with the so-called peace of the inner self. The latter is the refuge we so readily withdraw to. This can never be satisfactory. Loy points out that the issue is structural as well as personal, making the challenge that of changing the economic and political systems rather than remaining in blissful denial. He identifies the two main obstacles as:

1. Changing the mind is where it's at – self-absorption in the separate self - the deal we fall into.

2. Beliefs of Buddhist practitioners that we do not waste time trying to reform the unsatisfactory world, just concentrate on transcending it.

Both obstacles are major dharma mistakes, traps about higher spiritual reality that reflect disconnect in modern times, preventing us from engaging fully with the trainings and the world. Social, political and ecological engagements are devalued as we place our backsides on the cushion, chant, drink tea and avoid the reality around us. Modern Buddhism in the West definitely needs a wake-up call. The basic premise of the Bodhisattva Path is to walk it, not as a separate self, but as an engaged self.

Then an authentic sense of awakening naturally extends into political, economic and ecological spheres of potential action. I agree with David Loy that the reconstruction of our mind necessarily involves the reconstruction of our world – economic, political and spiritual.

I like his comment that "Bodhisattvas have a double practice – as they deconstruct and reconstruct, they also work for social and ecological change…….Such concerns are not distractions from our personal practice but deeper manifestations of it."

Thich Nhat Hanh was able to overcome this awkward divide when he created the Order of Interbeing during the Vietnam War. Socially Engaged Buddhism was renewed in Vietnam by him and then extended to the West. Thich Nhat Hanh ordained the first six members of the Order of Interbeing in February 1966 during the Vietnam War. The Order's foundation ethics for engaging with the wider society are the Fourteen Mindfulness Trainings created by Thich Nhat Hanh. They contain the Five Mindfulness Trainings, the Noble Eightfold Path and are a renewal of the earlier Bodhisattva Precepts. Thich Nhat Hanh was up to date and in tune with our times. He ensured the Fourteen Trainings of the Order are in step with modern historical, cultural and socio-economic developments yet rest on the foundation provided by the Buddha and 4th century expressions of socially engaged Buddhism.

Thich Nhat Hanh's book *Lotus in a Sea of Fire* and the fourteen ethical statements that he carefully sculpted, presented a revolutionary statement of Engaged Buddhism. Since 1966, the revolutionary part has been diluted, particularly in the West where the disconnect noted by Loy is in full swing. The Order of Interbeing established by Thich Nhat Hanh seems in the twenty-rst century to have lost the revolutionary fervour.

To emphasize that it is not just me who is way out on a limb here, I refer to a senior Theravada monk and scholar – Bhikkhu Bodhi (Buddhadharma Spring 2017). This respected monk looked at Donald Trump's "cabinet of bigotry" and at the same time noticed the absence of Buddhists on a petition of objection to it, which was signed by 2,500 religious leaders in the United States. He asked the obvious question; "why are Buddhists not visible as advocates for peace, sanity and social justice?' Where are they indeed, given that Buddhism is the pre-eminent religion of peace and compassion? He stated forcibly that not to participate in active engagement with politics, environmental and worldly events runs counter to the Buddha path of enlightenment. He points out that Buddhists fail to realize that the battleground over power and position are ethical contests. Trump's ascendancy to power shakes every Buddhist Mindfulness Training and this requires a strong push back from Buddhist leaders. So where is our agenda of collective resistance?

Bhikkhu Bodhi urges Buddhist advocacy in alliance with progressive leaders – religious and lay – to defend the United States' embattled democracy and leads the charge of relating to the trainings in a way that has no disconnect with present global concerns. That is the point of this essay – for there is nothing wrong with the trainings, apart from some essential rewording. The disconnect lies with contemporary Buddhists in the West who do not engage with the intent of the Trainings laid out by the Buddha and Thich Nhat Hanh. The Trainings are right here! Do we engage with them from the vantage points of self-seeking and separate-self OR engage with them from an open and engaged heart?

Bhikkhu Bhodi struck a chord with Buddhist leaders in the United States. I quote from an article in the May 2017 edition of the Lion's Roar magazine:

"Thirteen leading Buddhist teachers, joined by over 200 additional signatories, called on Buddhists and all peoples of faith to take a stand against policies of the new United States administration that will create suffering for the most vulnerable in society… Feeling the reality of this suffering, we remember that peacefulness does not mean passiveness and non-attachment does not mean non-engagement… The dharma is not an excuse to turn away from the suffering of the world, nor is it a sedative to get us comfortably through painful times. It (the dharma) is a powerful teaching that frees and strengthens us to work diligently for the liberation of beings from suffering…..While Buddhism has traditionally emphasized the personal cause of suffering, today we also discern how the three poisons of greed, aggression, and indifference operate through political, economic and social systems to cause suffering on a vast scale…"

As we resist the heightened threat of many of the new administration's policies, we also recognize that under-represented and oppressed communities in the United States have long suffered from systemic greed, aggression, aversion and indifference…….While some argue that the principle of non-duality suggests that Buddhists should not engage in or take sides on political or social issues, we believe the opposite is true. It is because we and others are not separate that we must act……..It is true that our numbers are small, yet we can join with others who share our convictions and values. For those who are new to this, please remember that there are many people who have dedicated their lives to the work of social change. They have the useful skills of compassionate organizing and building sustainable movements. Find them, get involved and learn from them." This May 2017 Manifesto is a major step in relieving the disconnect problem in Buddhism. This now brings me to the tricky role of Impermanence.

Impermanence

To change structures of elitism, greed and corporate dominance requires a mass change in consciousness. Mindfulness supports this change. The Buddha's teachings on impermanence also spur such a radical change. Can we grasp the insight of extinction – of ourselves, our civilization – even of the planet?

Without the insight of impermanence, we will not be able to change our mindsets. We have to find a way to adjust to our changed political and environmental circumstances. We can no longer hold on to a view of how it once was. Once we can accept that we have created the present global situation, then and only then can we find a respite, discovering insights that bring radical change to our values, habits and mindset.

It is very difficult in our western culture to accept death. The usual response is fear and denial. We have to re-educate our minds to get past these two obstacles. When we recognize our present form of civilization is dying, we will recognize despair and denial will do us no good. We need only find the courage to surrender and rely on our practice of mindfulness to provide a measure of safety. Instead of denial, space opens in our mind for lucidity and steadiness to enter, which could propel our species to live differently.

Such a future on Earth requires a mass awakening of attributes that run counter to the ecology of greed. It requires a candid acceptance that our global civilization in its present form is coming to an end. Such an acceptance of our true reality on the planet can alleviate the course of environmental collapse. The energy and power to avert the disaster facing us rests in our minds and in a new collective choice to live very differently. Thich Nhat Hanh brings this home to us in a direct and challenging way, making it very clear that any view not based on impermanence is wrong.

He shows how the Buddha provided meditations on impermanence for his followers so they could recognize the only thing that follows death is the fruit of our action and thinking, of our speech and of our acts during our lifetime. Specifically, on climate change he's very blunt:

> "If we continue to consume unwisely, if we don't care about protecting this wonderful planet….the ecosystem will be destroyed to a large extent and we will need millions of years to start a new civilization. Everything is impermanent…. We are our environment, which is in a process of self-destruction."

This brings a certain peace and clarity to our minds and perhaps we can implement ethics, structures and technology to ensure a niche on this planet. We have a job to do in terms of cultivating a transformation in our consciousness, bringing about a new way of living in harmony with one another and on Earth.

We must deliberately cultivate positive ethical attributes in our minds. We have to shine the light of recognition and mindfulness on our suffering, so we can become steady and full of resolve to live differently. We have to shift the tide of negativity, change our mindset and not squander our life. The Five and Fourteen Mindfulness Trainings provide us with templates to do that, as we consciously choose to nurture patterns of behaviour and habits that are wholesome and generous. In other words, we make mindfulness practice our new habit. It is an internal transformation of consciousness at the core of our being.

I shape all of this into a simple personal mantra: "I refrain from causing harm." I know by refraining from one thing that causes harm, I then prevent other harmful things from happening. I arrive at my own insight, not imposed by outside authority and know it takes mindfulness to do this.

The Five Mindfulness Trainings provide the starting point, a guidance system and a deep well of internal ethics to live by. My commitment is to actualize these trainings in my life, and in the lives of others, to the best of my ability.

I issue a Call to Action and bring Bhikkhu Bodhi back. In Buddhadharma, spring 2017 he urges Buddhist advocacy in alliance with progressive leaders to defend the United States' embattled democracy from President Trump's "cabinet of bigotry."

He states; "We can call in unison for a policy of global generosity in place of rash militarism, for programs that protect the poor and vulnerable, for the advancement of social and racial justice, and for the rapid transition to a clean-energy economy ....and bring the moral weight of the dharma to bear on matters that affect the lives of people anywhere – now and long into the future." His statement was followed by the stance taken by Buddhist leaders in the May 2017 issue of Lion's Roar magazine

I also call out the Hopi Elders' Prophecy in 2000:

"Create your community. Be good to one another. And do not look outside yourself for your leader… See who is there with you and celebrate…. All that we do now must be done in a sacred manner and in celebration. We are the ones we've been waiting for."

For our part we can work with municipalities, conservationists and River Keepers to clean up our waterways and environment. Ensure that children in schools go with you and prepare them to handle cyberbullying and neglect. We hold politicians and corporations to account. Create coalitions with progressive organizations who share our love of kindness and decency.

Walk upon the Earth – Lightly. Be fully Here and Present – Lightly.

EPILOGUE: **The Merchant and the Diamond**

There was a merchant who lived in a far-away land. He was very wealthy and built a trading empire that brought him great riches. He was respected throughout the land for his fairness and astuteness, yet in the midst of all his wealth and fame he felt a lack, that there was something he did not have. He did not know what it was.

One night he had a dream and remembered it very clearly. He dreamed there was a monk sitting under a tree at the forest's edge, and that this monk had something special to give to him. He was not accustomed to dreaming, so he felt this dream held a special portent.

At sunrise the next day he left his house and walked to the edge of the town where he lived. He saw the monk, just as in his dream, wearing a saffron robe, sitting quietly under the shade of a tree at the edge of the forest. He looked so peaceful.

As the merchant approached, the monk opened his eyes and smiled gently to him. The merchant stopped, bowing respectfully, and said:

"Dear monk, I had a dream about you last night, that you would be sitting here on the edge of the forest and that you had something for me. What is it that you have for me?"

The monk paused for a moment then slowly reached into his canvas bag. He brought out a huge diamond as big as a man's fist. It sparkled and shone in the sunlight, dazzling the eyes and senses of the merchant. It was the most beautiful and valuable diamond the merchant had ever seen. The monk said, "Have you come for this?"

The merchant without hesitating replied, "Oh yes, thank you so much, this is wonderful. I have always wanted to possess such a magnificent diamond." He thanked the monk profusely for this unexpected and magnificent gift, wrapped the diamond inside his jacket and returned home.

Once there he placed the diamond in his front room. He closed the shutters and locked the doors to his house and stayed with the diamond, totally mesmerized and entranced by its beauty and purity. He did not go to work that day, nor did he eat or drink. He thought that this gift was the missing piece of his life and he wanted to bask in the glory of it. When he went to sleep that night, he placed the diamond on a small pedestal by his bedside, so that he could have it close by. Yet he could not sleep. He felt a strange disturbance within himself that he did not understand. He tossed and turned, not knowing what to do about the growing restlessness. Just before sunrise he rose, got dressed and carefully wrapped the diamond in a cloth, before setting off once more for the edge of the forest.

The monk was sitting in the same place, deep in meditation. The peace that emanated from him calmed the restlessness that so disturbed the merchant. On hearing the merchant's footsteps come closer and then stop before him, the monk opened his eyes and once more smiled very gently to the merchant. "Good morning my friend," he said. "Are you not happy with the diamond?"

The merchant bowed and placed the diamond at the feet of the monk and said, "Good monk, it is not the diamond that I want. I would like to have the heart that can give away such a diamond."

The monk very quietly stood up and bowed to the merchant, "Good sir, as that is your wish, I will teach you to meditate."

Postscript for Our World is Burning

My life as a global traveller stretched my attention beyond any limits I could have placed on it. From teenage volunteering in Borneo with Britain's Peace Corps to being a yogi in India and a Zen teacher in Canada – expansion of mind was inescapable.

My challenging journey navigates shamanic healing of childhood sexual abuse, guru training as well as a near death experience in an ashram in India. In my teenage expedition to Sarawak, Borneo, I kept a journal of the vivid surrounds. From that time on I kept journals and scribbled poetry. I eventually had trunks full of journals that reminded me of the experiences I went through, particularly in India.

My extensive shamanic training with First Nations medicine people was also carefully logged. As an anthropologist I was fortunate to encounter many story tellers across North America – Dene, Hopi, Ojibwa, Algonquin, Inuit – to mention a few. Their poetic recounting of myths and history had a deep impact upon me. I would say that without poetry, cultures implode.

Over a period of thirty years, four extraordinary medicine people enhanced my process of remembering the power of the poetic voice. Through their mentoring I learned how to reconfigure my understanding of time, place and consciousness. I also chose to listen to the feminine voice of Earth Wisdom rather than the multitude of competing voices in my deep unconscious. This impact shows up in my writing.

From this vast range of experience I found an ability to sculpt narrative in a novel way and this was expressed in my poetry and books.

I certainly stumbled through the first part of life, but then stood strong in my own sovereignty in the latter part. In India, Arizona, France and Canada's wilderness, I went to great lengths to transform karma and began to live life as a Meditation for Gaia. My journal notes were a signpost to always be authentic, even when it was difficult to re-read. As a professor I wrote text books and scholarly papers, which had particular protocols that were often stifling. When I started late on the writing craft – I had to re-learn how to write without sounding pompous, to be mindful of what the reader could take in. I gave up on footnotes.

I am a Zen teacher, also a recognized guru in India. My initial task is to refine my own consciousness - to be a vehicle to chart an authentic path. If I did not do this, then I could not write the way I do. The focus on daily mindfulness from my Zen practice enables me to be still and clear. From this energy the poems and chapters emerge. I do my best not to write from a space of frustration or of wanting to get the writing finished. I wait until the energy of mindfulness is tangible – then creating the words and text just flow.

My books are epic tales that seamlessly weave together to create inspiration for a wide range of fellow spiritual seekers, environmentalists, Generation X and Y, feminists, students and academics alike. I recognized early on that global citizens are staring into the abyss – yet instead of being eaten up by it all, I say to them: "Awaken Spiritually," for that transforms everything. We have made our world an unpredictable beast because we fail to work with it intelligently. We have to take back control of ourselves and this is a spiritual matter. Turning on the switch of awakening seems to be a good idea!

That is the prod and direction of my poems and books. We just need to touch the sacred in ordinary experiences of daily life to find the courage and determination to transform.

My writing caps a long-term fascination with consciousness. As a Professor I taught courses on Ecology, Symbols, Engaged Buddhism and Meditation Systems. I am a healer, mentor and educator, able to encourage people through example to find their true nature so that humanity and the world may be renewed. All of this funnels back into my writing.

My hope is that the reader will summon the courage to believe that they can take steps to transform internally and then make a difference externally. The stories I tell are offered as a gift for our planet. My purpose is to share my wealth of experience on how to live in harmony not just with ourselves but with the place we call home… Earth.

I shed light on issues that will affect our world for generations to come. The example of my own challenging journey and personal transformation illuminates a path for others to expand their consciousness and chart the course for a future beyond the abyss. The human race does not need to be stuck with maladaptive options and patterns. We can and must transform.

My Essays deliver a vigorous message about personal transformation in order to become different stewards of the earth and society. In the Sixteen Essays, I offer reality-based information that is in high demand in today's society, which provides the potential for my projects to become fresh, new icons for today's hungry culture. Hungry, that is, for authentic transformation. Authenticity is the bottom line, which took a while to register in my career as an academic. It is essential to find one's authentic voice and not be afraid to express it.

In Chapter One my grand-nephew James requested a training program. I offered this 8-point plan.:

1. You – learn to be silent and quiet! Clear time and space for spiritual practice at home and throughout your daily schedule.

2. Create a stress reduction menu and subtract the "weeds" in the garden of your mind.

3. Be determined to meditate daily – do the weeding.

4. Focus on and soften your heart – nurture the soil of your mind's garden.

5. Cultivate the seeds of mindfulness at home, school, and work or in solitude.

6. Simplify, make do with less, de-clutter your mind and home.

7. Taste the fruits of your spiritual practice that change your mind.

8. Engage with the world.

This plan for James, or something like it, is a necessary prelude for mindful engagement. Whether it is at home, in community, activism against damaging global structures – some form of reflective quiet enables citizens to connect, reach out and play a part in reconstructing our place on planet earth. To stay on the sidelines doing nothing, creates victims of us all.

There is no time to lose. I remember saying to James that "The greatest gift we can give to ourselves and others at this time of global crises is sharing and caring. It involves stepping onto what the Buddhists call the Bodhisattva Path." I explained that a Bodhisattva was a person who stayed in the global mess and did their best to awaken the minds and hearts of people. I firmly stated that it is time for the Bodhisattva-within-us to enter the 21st century as the example for action.

It takes training, practice, intelligence and creative vision to find the drive to create a tangible spirit of co-operation, the willingness to share and be supportive, and learning how to cross the bridges of conflict.

This thread of understanding finds a place in every essay in *Our World is Burning*. However, the obstacles preventing people taking wise action are a mixture of fear, despair, disempowerment and a sense of hopelessness. The overwhelming terror of Gaia's collapse, along with the consequences of global refugees and fascist regimes are unbearable. Our challenge is to be in society, but as a still island of mindful engagement. We do not have to be caught by pathological consumerism. Voluntary Simplicity is a good starting place, becoming aware of our consumerism. We can also participate and engage in global protests through organizations such as Avaaz. We take action and get up close and personal with the crises, reduce our ecological footprint, bring ethics into business and the workplace, support science and diversity, and warn governments and corporations.

Where do we start? Of course we must think globally and be aware of the bigger picture and step beyond the smaller pictures of ourselves created from fear and disempowerment. Yet we can also act locally with great vigour in our families and communities. Our intentions then spread as ripples from a pebble dropped in still water.

In addition to holding officials, politicians and corporate culture to account let us begin with the small things that all of us can do. While at the same time alerting the political and corporate decision makers that we do mean business as voters and consumers deeply concerned about the planet and our location on it. This is very important. Our leaders are a manifestation of our collective will. When the collective will changes, our leaders will act differently.

A massive global citizen response will certainly elicit an equally massive government and corporate response, as the bottom-up movement and top-down strategies for drastic change meet and integrate.

There is no room in a Global Ecological Emergency for separating into "US' and "THEM" categories.

We are interconnected whether we like it or not. We will all live together or we will all die together.

An intelligent and all-encompassing green ideology embedded in everything we produce and market is a way to bridge competing agendas.

Our dependence on fossil fuels will then reduce because we are aware of the deadly consequences of our addiction to oil and coal.

The transition to a carbon neutral global energy system over the next few decades will be costly and require a massive response from government and corporate leaders to initiate a new industrial revolution. This is necessary to blunt the impact of climate change. It is a huge global industrial project that governments and corporations can bring about due to citizen pressure to "Make It So!"

Climate Change has certainly entered public consciousness. It just has to penetrate the corridors of political and corporate power.

As global citizens we must find ways and means to support the shift in consciousness at all levels of global society to make this so.

Our future existence, and the existence of other species on planet earth, depends on your making a new beginning for all of us.

About The Author

Dr. Ian Prattis is Professor Emeritus at Carleton University in Ottawa, Canada. He is an award winning author of fifteen books. Recent awards include Gold for *Redemption* at the 2015 Florida Book Festival, 2015 Quill Award from Focus on Women Magazine for *Trailing Sky Six Feathers* and Silver for Environment from the 2014 Living Now Literary Awards for *Failsafe; Saving the Earth From Ourselves*. His novel – *Redemption* – is being made into a movie. His poetry, memoirs, fiction, articles, blogs and podcasts appear in a wide range of venues.

He was born in the UK and has spent much of his life living and teaching in Canada.

His moving, eye-opening books are a memorable experience for anyone who enjoys reading about primordial tendencies. Beneath the polished urban facade remains a part of human nature that few want to acknowledge, either due to fear or simply because it is easier to deny the basic instincts that have kept us alive on an unforgiving earth. Prattis bravely goes there in his outstanding literary work.

A Poet, Global Traveler, Founder of Friends for Peace, Guru in India, and Spiritual Warrior for planetary care, peace and social justice. A Zen teacher, Ian presently lives in Ottawa, Canada and encourages people to find their true nature, so that humanity and the planet may be renewed. He mostly stays local to help turn the tide in his home city so that good things begin to happen spontaneously.

Born on October 16, 1942, in Great Britain, Ian grew up in Corby, a tough steel town populated by Scots in the heartland of England's countryside. Cultural interface was an early and continuing influence.

Ian was an outstanding athlete and scholar at school, graduating with distinctions in all subjects. He did not stay to collect graduating honours, as at seventeen years old he travelled to Sarawak, Borneo, with Voluntary Service Overseas (1960 – 62), Britain's Peace Corps.

He loved the immersion in the myriad cultures of Sarawak and was greatly amused by the British colonial mentality, which he did not share. He worked in a variety of youth programs as a community development officer, and also explored the headwaters of Sarawak's major rivers, with expeditions into Indonesian Borneo.

Returning to Great Britain after Sarawak was an uneasy transition. He did, however, manage to get through an undergraduate degree in anthropology at University College, London (1962 – 65), before continuing with graduate studies at Balliol College, Oxford (1965 – 67).

At Oxford, academics took a back seat to the judo dojo, rugby field, bridge table, and the founding of irreverent societies at Balliol. Yet by the time he pursued doctoral studies at the University of British Columbia (1967 – 70), his brain had switched on.

He renewed his passion for other cultures, placing his research on North West Coast fishing communities within a mathematical, experimental domain that the discipline of anthropology was not quite ready for. Being at the edge of new endeavours was natural to him, and continues to be so.

He has been a Professor of Anthropology and Religion at Carleton University since 1970. He has worked with diverse groups all over the world and has a passion for doing anthropology. "It's better than having a real job," he says, "everything changes and the only limits are your imagination and self-discipline."

His career trajectory has curved through mathematical models, development studies, hermeneutics, poetics and symbolic anthropology, to new science and consciousness studies.

The intent was always to expand, and then to cross, existing boundaries; to renew the freshness of the anthropological endeavour, and make the discipline relevant to the individuals and cultures it touches.

He studied Tibetan Buddhism with Lama Tarchin in the early 1980's, Engaged Buddhism with Zen Master Thich Nhat Hanh much later, Christian meditation with the Benedictines, and was trained by Native American medicine people and shamans in their healing practices. He also studied the Vedic tradition of Siddha Samadhi Yoga, and taught this tradition of meditation in India. He was ordained as a teacher and initiator – the first westerner to receive this privilege – acknowledged in India as a guru.

Later in life, as a respite, he lived in a hermitage in Kingsmere, Quebec, in the middle of Gatineau Park forest when his pet wolf was alive.

He facilitated a meditation community in Ottawa called the Pine Gate Mindfulness Community from 1977-2017. At the outbreak of the Iraq war he founded Friends for Peace Canada – a coalition of meditation, peace, activist and environmental groups to work for peace, planetary care and social justice. He is also the editor of an online Buddhist Journal.

Since retiring from Carleton University in 2007 he has authored four books on dharma, two on the environment, a trilogy titled "Chronicles of Awakening" and this collection of essays. He enjoys the freedom to create at his own pace. He received the 2011 Ottawa Earth Day Environment Award – and he has yet to discern the ordinary meaning of retirement!

Ian Prattis / *Our World is Burning*

Manor House Publishng
www.manor-house.publishing.com
905-648-2193

www.ingramcontent.com/pod-product-compliance
Lightning Source LLC
Chambersburg PA
CBHW050222270326
41914CB00003BA/524